MATHS & ENGLISH FOR
ELECTRICAL

Graduated exercises and practice exam

Andrew Spencer and Robert Henley

CENGAGE
Learning·

Australia • Brazil • Japan • Korea • Mexico • Singapore • Spain • United Kingdom • United States

Maths & English for Electrical

Andrew Spencer and Robert Henley

Publishing Director: Linden Harris

Commissioning Editor: Lucy Mills

Development Editor: Claire Napoli

Production Editor: Alison Cooke

Production Controller: Eyvett Davis

Marketing Manager: Lauren Mottram

Typesetter: Cenveo Publisher Services

Cover design: HCT Creative

For product information and technology assistance, contact **emea.info@cengage.com**.

For permission to use material from this text or product, and for permission queries, email **emea.permissions@cengage.com**.

This work is Adapted from Pre Apprenticeship: Maths & Literacy Series by Andrew Spencer, published by Cengage Learning Australia Pty Limited © 2010.

British Library Cataloguing-in-Publication Data
A catalogue record for this book is available from the British Library.

ISBN: 978-1-4080-7753-5

Cengage Learning EMEA
Cheriton House, North Way, Andover, Hampshire, SP10 5BE, United Kingdom

Cengage Learning products are represented in Canada by Nelson Education Ltd.

For your lifelong learning solutions, visit
www.cengage.co.uk

Purchase your next print book, e-book or e-chapter at
www.cengagebrain.com

Printed in Malta by Melita Press
1 2 3 4 5 6 7 8 9 10 – 15 14 13

Maths & English for Electrical

Contents

Introduction

It has always been important to understand, from a teacher's perspective, the nature of the maths skills students need for their future, rather than teaching them textbook mathematics. This has been a guiding principle behind the development of the content in this workbook. To teach maths and English that is *relevant* to students seeking apprenticeships is the best that we can do, to give students an education in the field they would like to work in.

The content in this resource is aimed at the level that is needed for a student to have the best possibility of improving their mathematical and English skills specifically for Electrotechnical and Electrical Installation. Students can use this workbook to prepare for their functional skills assessment, or even to assist with basic maths and English for their Electrotechnical qualification. This resource has the potential to improve students' understanding of basic mathematical concepts that can be applied to the Electrical industry and Electrotechnical environment. These resources have been trialled, and they work.

Commonly used industry terms are introduced so that students have a basic understanding of terminology they will encounter in the workplace environment. Students who can complete this workbook and reach a higher outcome in all topics will have achieved the goal of this resource.

The content in this workbook is the first step to bridging the gap between what has been learned in previous years, and what needs to be remembered and re-learned for use in exams and in the workplace. Students will significantly benefit from the consolidation of the basic maths and English concepts.

In many ways, it is a 'win-win situation', with students enjoying and studying relevant maths and English for work and training organizations and employers receiving students that have improved basic maths and English skills.

All that is needed is patience, hard work, a positive attitude, a belief in yourself that you can do it and a desire to achieve. The rest is up to you.

About the authors

Andrew Spencer has studied education both within Australia and overseas. He has a Bachelor of Education, as well as a Masters of Science in which he specialized in teacher education. Andrew has extensive experience in teaching secondary mathematics throughout New South Wales and South Australia for well over 15 years. He has taught a range of subject areas, including Maths, English, Science, Classics, Physical Education and Technical Studies. His sense of the importance of practical mathematics has continued to develop with the range of subject areas he has taught in.

Robert Henley has adapted Maths & English for Electrical. Robert, 'Bob' started his career as an apprentice vehicle electrician at the Hants and Dorset bus company. A second 'apprenticeship' in electrical installation followed and then an honours degree in electrical and electronic engineering. Following graduation from the University of Portsmouth at the age of forty five, Bob has been teaching electrical engineering subjects in the post-sixteen sector. Bob has also achieved a post graduate certificate in post-compulsory education.

Acknowledgements

Andrew Spencer:
For Paula, Zach, Katelyn, Mum and Dad.
   Many thanks to Mal Aubrey (GTA) and all training organizations for their input.
   To the De La Salle Brothers for their selfless work with all students.
   Thanks also to Dr Pauline Carter for her unwavering support of all maths teachers.
   This is for all students who value learning, who are willing to work hard and who have character … and are characters!

Robert Henley: For Sharon my wife, friend and life anchor.
For Emma and Clare.
For Aimie, Alan, Robert, Eliza and William.
   Many thanks to Mrs. Gannaway, who taught me the secrets of punctuation when I was ten years old, (and helped me realize that I wasn't stupid after all).
   In Memory of Richard Jones. Electrician, teacher, music buff and collector of all sorts of things. Peace be with you my friend.

ENGLISH

Unit 1: Spelling

Short-answer questions

Instructions to students

- This exercise will help you to identify and correct spelling errors.
- Read the following question and then answer accordingly.

Read the following passage and identify and correct the spelling errors.

An electrician begins work on a new two-storey construction. The toolkit includes a muldimeter, a set of plyers, tape mesure, spaners, a set of alen keys and various screwdrivas. The aprentice turns up at 8.00 a.m. with the work van. They both start work on the grownd floor. The walls need driling, so that the first set of wiares can be threeded through. The electrishan uses the 18 V drill. The battary was fully charged and did the job easerly. The apprentice goes to the work van, gets the 8 m extention lead and attaches the hammar drill. He puts in a 10 mm drill bit and tightans the drill with the chuck. Four holes are driled and the main wires are put through. The ends of the wires are krimped as they finish off before having a break.

Incorrect words:

_____

_____

Correct words:

_____

_____

Unit 2: Comprehension

Section A: Reading skills

Short-answer questions

Instructions to students

- This is an exercise to help you understand what you read.
- Read the following passage and then answer the questions that follow.

Read the following passage and answer the questions in sentence form.

Mike the electrician started work early on Friday morning. He arrived at the building site at 5.45 a.m. Adam, who is also a 'sparky', arrived 45 minutes later. A new security system needed to be put in and Mike started unloading the work van. Adam grabbed the toolbox and checked that the cordless drills were fully charged. Mike brought out the CCTV (closed circuit TV) components and both the electricians began to set up the job.

Once the scaffold tower was in place, Mike took the cordless drill, loaded up some of the equipment and began to climb the scaffold tower. He established the position that he wanted to drill the first hole, and then began positioning the first components of the CCTV. Pilot holes needed to be drilled first in order to allow the larger holes to be accurately drilled. Mike had completed similar jobs like this one numerous times. Meanwhile, Adam was uncoiling the wire necessary to connect the CCTV to the main electrical system. The circuit needed to be wired through the junction box and the surge arrester.

Mike found that there was no problem securing all of the components due to the pilot holes he had drilled. Adam had connected up the electrical circuit without a problem and a test showed everything was working fine. Mike and Adam had a lunch break at 12.15 p.m. for an hour and then returned to complete the last of the work. They both finished working at 4.10 p.m.

QUESTION 1

How many hours and minutes did Adam work for the whole day?

Answer:

QUESTION 2

What does the acronym CCTV stand for?

Answer:

QUESTION 3

Why did Mike drill pilot holes?

Answer:

QUESTION 4

What was Adam going to use the cable that he had uncoiled for?

Answer:

QUESTION 5

What did Adam and Mike need to do before they began working to know that their equipment was safe to use?

Answer:

Section B: Comparing different types of text

Short-answer questions

Specific instructions to students

- This is an exercise to help you identify different types text.
- Read the activity below, then answer accordingly.

Read each of the following paragraphs, state the purpose of each type of text, explain whether the text is formal or informal and why it is appropriate in this context.

Text A – Dave, can you pick up the shower unit from the Electroparts and then go round to Mrs Jones in Valley road to fix her lights. Thanx Mo.

PURPOSE OF TEXT:

_____

_____

FORMAL/INFORMAL:

_____

WHY THE TEXT IS APPROPRIATE:

_____

_____

_____

Text B – To isolate the lighting circuit turn on all the lights in the house and isolate each circuit in turn at the fuse box. When the lights go out on the correct circuit, you know which fuse is controlling that circuit.

PURPOSE OF TEXT:

_____

_____

FORMAL/INFORMAL:

_____

WHY THE TEXT IS APPROPRIATE:

_____

_____

_____

Text C – Please be advised that should your account not be paid in full, within 30 days of this notice, our collections team will be instructed to take legal proceedings to recover all of your outstanding debt.

PURPOSE OF TEXT:

_____

_____

FORMAL/INFORMAL:

_____

WHY THE TEXT IS APPROPRIATE:

_____

_____

_____

Text D – Work carried out:

- 1 double socket and pattress changed
- 2 way lighting repaired
- All work tested and certificated

PURPOSE OF TEXT:

_____

_____

FORMAL/INFORMAL:

_____

WHY THE TEXT IS APPROPRIATE:

_____

_____

_____

Section C: Factual and subjective text

Short-answer questions

Specific instructions to students

- This is an exercise to help you identify factual and subjective text.
- Read the activity below, then answer the questions accordingly.

Text can either be factual or subjective.

If a piece of text is factual it is based on real evidence and records of events which is not biased by the writer's opinion.

If a piece of text is subjective, it contains an individual's personal perspective, feelings or opinions which might differ from another individual's view of the same subject.

Which of the following is fact and which is opinion?

QUESTION 1
The closing ceremony of the 2012 Olympics was the best that I have ever seen.

QUESTION 2
Bradley Wiggins won the tour de France in 2012.

QUESTION 3
LED lighting systems are efficient.

QUESTION 4
If low voltage hand tools are used on a worksite they must be protected by an RCD.

QUESTION 5
DeWalt make an extensive range of power tools for the construction trades.

QUESTION 6
You must wear the appropriate PPE when working on site.

QUESTION 7
The letters IET stand for Institute of Engineering Technology.

QUESTION 8
Hotdogs are much nicer than burgers for lunch.

QUESTION 9
My local electrical store sells the best quality equipment.

QUESTION 10
Crabtree make accessories for the electrical trade.

Unit 3: Grammar

Section A: Apostrophes

Short-answer questions

Instructions to students

- This exercise will help you to use apostrophes correctly.
- Read the following question and then answer accordingly.

Using apostrophes can be confusing, but with some 'know-how', it becomes a little easier. You just need to know what an apostrophe is and a few basic rules on how and when to use them.

There are two main reasons for using apostrophes.

Reason 1: To show a letter, or letters, have been left out in a word, or letters that are left out when two words are joined together. These are sometimes called contractions in grammar. **Contractions are <u>not</u> usually allowed in technical reports and essays, but are often found in stories and magazine articles.**

EXAMPLE

Cannot	-	Can't
Do not	-	Don't
I am	-	I'm
Let us	-	Let's
They are	-	They're
They will	-	They'll
Will not	-	Won't
You will	-	You'll

Insert the appropriate contraction in the following sentences.

1. (We are) _____ due to be at the construction site at 7.30 a.m. on Tuesday morning.

2. (It is) _____ considered good manners to wipe your feet before you enter a customer's house.

3. (I am) _____ working hard to be the best electrician that I can possibly be.

4. (Do not) _____ work on live circuits, (it is) _____ a very dangerous practice.

5. Study hard at work, college and home, (it will) _____ be worth it when (you have) _____ qualified as an electrician and are earning a good wage.

Reason 2: To show possession. This is used when something belongs to someone or a group. This can be a tangible physical/touchable/material item, such as a mobile phone, or something that is not, such as an emotion or an opinion. When we use an apostrophe to show possession, the apostrophe has an 's' attached to it, <u>unless</u> the word that we are showing as the owner already ends in an 's', in which case we only attach an apostrophe.

Here are some examples to make it a bit easier to understand.

- Simon owns a book   -   _**Simon's**_ book.
- David has a calculator   -   _**David's**_ calculator.
- The electrician has an apprentice called Mohamed   -   The _**electrician's**_ apprentice is called Mohamed.
- Thomas owns an electrical maintenance company called 'Tomlec'   -   _**Thomas'**_ company is called 'Tomlec'.

Place the apostrophes in the correct places in the following sentences.

a. 'Well that was a long shift' said Adam as he came up to the end of Billys road. 'ay what' came the response from Billy as he pulled the earbuds from his MP3 from his ears. 'Sorry Adz ... was listening to...' Adam cut him off 'you were nodding off' he teased.

b. Billy was Adams apprentice and they had been working together for about 18 months.

c. They both worked for Harris Electrical, a small electrical contracting company based in the south midlands. Adam had been one of Harris first apprentices and had worked with old man Harris himself.

d. Mr Harris, as Billy referred to the boss when he spoke to him, had built the business from scratch and now employed ten electricians and three apprentices. Mr Harris was also Billys dads best friend.

e. 'Get yourself a good nights sleep lad', Adam instructed as he pulled up outside of Billys house. 'We have a busy day tomorrow at Wilsons factory, so Ill pick you up at 7, dont be late.'

Section B: Similar sounding words

Short-answer questions

Instructions to students

- This exercise will help you to identify and correctly use similarly sounding words.
- Read the following question and then answer accordingly.

There, their, they're.

These words look and sound similar and it can be challenging to decide how to use there, their and they're correctly. Read the information in each of the examples, think about them carefully, then attempt the paragraph below and you should be able to complete it easily.

There indicates a place, e.g. put the cable over *there* please, or *there* is the van.

Their indicates a belonging, e.g. *the electricians stored their tools in the site office*, or *customers like to have their work completed on time*.

They're has an apostrophe in it, it indicates that something has been left out. In this case it is the letter 'a' and is a contraction of 'they (a)re'.

Complete the following paragraph by inserting the appropriate there, their or they're.

Staying safe at work is very important and _____ are lots of things to watch out for. _____ are many things in the workplace that can hurt you and individuals must do all that they can to maintain _____ own safety. Safety is so important that _____ are laws that make it the responsibility of every individual to take care of _____ own safety; the law also says that _____ responsible for the safety of the people around them.

Wear, where, were, we're.

These words look and sound similar and it can be challenging to decide how to use wear, where, were and we're correctly. Read the information in each of the examples, think about them carefully, then attempt the paragraph below and you should be able to complete it easily.

Wear has two meanings:
1. To have something on your body e.g. *you need to wear safety goggles when using a drill*, or *it's important to wear a hard-hat when there is a risk of falling objects*.
2. To have an object eroded by friction e.g. *when motor bearings become dry, they tend to wear out*.

Where indicates a place,
e.g. this is *where I have put the cable*, or *where is the van?*

Were indicates that something was done in the past,
e.g. *the electrician's tools were stored in the site office*, or *the jobs were completed on time*.

We're has an apostrophe in it which indicates something has been left out. In this case it is the letter 'a' and is a contraction of 'we (a)re'.

Complete the following paragraph by inserting the appropriate wear, where, were or we're.

An important part of protecting yourself at work is to _____ the appropriate safety clothing and equipment. To protect your skin, it is recommended that, barrier cream is applied to your hands and that you _____ overalls to cover your arms and legs. _____ there are more specific hazards or dangers, such as noise, bright light, possible falling objects or moving vehicles then you will possibly need to _____ ear defenders, eye protection, a hard-hat or a high-viz jacket. If you _____ to be involved in an accident the consequences can be drastic. Remember that when we are in the workplace _____ all responsible for safety.

To, too, two.

These words look and sound similar so it can be challenging to decide how to use to, too and two correctly. Read the information in each of the examples, think about them carefully, then attempt the paragraph below and you should be able to complete it easily.

To has several meanings:
1. Indicating motion e.g. *the apprentice electrician is going to work*.
2. Approaching e.g. *by the end of the week, the installation team will be close to finishing the first house*.
3. Used as a joining word e.g. *the apprentice had a one-to-one interview with the manager*.

Too has two meanings:
1. Also e.g. *the apprentice stored his tools in the site office too*.
2. Used in relation to size e.g. *the apprentice's boots were too big for him*, or *the cable was too small*.

Two relates to the number two (2).

Complete the following paragraph by inserting the appropriate to, too or two.

Some jobs that electricians do require _____ people. Sometimes this is because there is _____ much work for one electrician, but more often the reason is that _____ people are required _____ be on site _____ maintain safety. Where there is a specific danger, such as working below ground, working at height or working on or near live circuits, it is considered best practice _____ have _____ people working as a team.

Unit 4: Punctuation

Short-answer questions

Instructions to students

- This exercise will help you to identify and correctly use punctuation in your written work.
- Read the following question and then answer accordingly.

Apostrophe – (') An apostrophe is used to show the contraction of two words into one. An apostrophe is placed where the letters have been dropped, e.g. 'I do not want to be late for work' becomes 'I don't want to be late for work'.

A possessive apostrophe is used to show possession or relationship, e.g. Jacek's screwdrivers, the apprentice's toolbox.

Brackets – ( ) Round brackets are used to add information for clarity, e.g. undo the three screws, (turn anti-clockwise), to remove the cover plate.

Colon – (:) and Semi-colon – (;) A colon is used to introduce a list, a quotation or a summary - e.g. Tools required: pliers, side-cutters and a range of screwdrivers.

A semi-colon is used to link two closely related sentences or statements, e.g. the site crew are due back on Thursday; they have been away for three days.

Comma – (,) A comma is used to mark a pause in a sentence, or used to separate items in a list, e.g. The customer required two socket outlets, three two-way switch plates and pattresses to suit.

Ellipsis – (...) This signifies place where something has been omitted or there is a pause or interruption. It is used for economy or style, e.g. He heard a strange noise from the boiler ... it wasn't a very encouraging sound!

Exclamation mark – (!) A punctuation mark is used at the end of a sentence to show great emotion such as surprise, anger or fear or to emphasize that something is important, e.g. DANGER 230 VOLTS!

Full stop – (.) A full stop is the usual end of a sentence.

Hyphen – (–) A hyphen, sometimes called a 'dash', may be used to replace brackets; to indicate an afterthought, to join two connected words or to replace other punctuation in informal writing - e.g. hard-hat or high-viz jacket.

Quotation marks – ("") and ('') Double quotation marks or speech marks are to indicate direct quotations, e.g. Bob replied "A bacon roll and a black coffee please."

Single quotation marks are used to indicate a definition or that a word or phrase has special meaning, e.g. a 'pattress' is a surface box for mounting electrical accessories.

The following passage is a section of work submitted by a trainee for marking; can you correct the spelling and punctuation?

Q. What is the purpose of the inspection process with regard to new electrical installations, and when is the inspection carried out?

The reson we do inspections is 2 make sure everyfing

is safe it complies wiv in the regs and the equipment is

suitable the inspection is done when it is installed and

befor u test it

Short-answer questions

Specific instructions to students

- The following questions will help you practise your grammar and punctuation.
- Read the following questions, then answer accordingly.

QUESTION 1

Which linking word or phrase could you use instead of 'whereas'?

Answer

_____

QUESTION 2

What does the linking word 'alternatively' mean?

Answer

_____

QUESTION 3

What punctuation is missing from the following sentence?

> We are main authorised dealers for Bosch Miele Neff and Siemens products and so naturally we supply only premium quality appliances.

Answer

_____

QUESTION 4

What is wrong with the following text? Correct the following sentences.

> Last Saturday Brilliant Electrical Ltd was extremely busy; grace was run off her feet with all the customers that were calling to book appointments for one of the company's electricians to fix their appliances. She was pleased that she had remembered to ask jim to help. She had felt a bit guilty as she had to send jim to a fix a dishwasher for a customer in manchester, which was 15 miles away from his previous job.

Answer

_____

_____

QUESTION 5

When you have completed a section of writing, what should you look for when checking through your work?

Answer

_____

_____

QUESTION 6

What is wrong with the following text?

> Why not visit our new electrical appliance showroom? Were a leading specialist supplier of top quality electrical brands. Our sales staff are fully trained in all appliances. We stock a large range of high quality kitchen appliances. We offer free delivery and an old appliance removal service. To find out more visit our shop on 8 Bridge Street, Cheedle or call 01435 778367.

Answer

_____

_____

QUESTION 7

Can you identify the mistake in this job application letter?

> Dear Madam
>
> I wish to apply for the vacancy of Trainee Electrician at Bright Electrics, as advertised in this week's Gloucester Globe.
>
> I have just completed my Level 2 NVQ Diploma in Electrical Installation course at Dinsdale Park Colleage and am now looking for work in the Gloucester area.
>
> I enclose a copy of my CV and look forward to hearing from you.
>
> Yours faithfully,
>
>
> Luke Smith

Answer

_____

QUESTION 8

Can you identify the mistake in this advert?

> **Solar panel installation for YOU!**
>
> Free solar electricity plus tax free earnings available with solar panel instalations provided by To The Sun - the market leading roof solar panel installer. Please ring 0151 629 4027 to find out more about our special services.

Answer

QUESTION 9

Add punctuation to the following text to make the sense clearer.

> Before we can begin we must first obtain permission to isolate the circuit this must be obtained from the person responsible for the electrical installation (the duty holder) not just any employee as we are going to be isolating the supply the duty holder must ensure that the safety of persons and the operation of the business is not going to be compromised

Answer

Short-answer questions

Specific instructions to students

- This is an exercise to help you understand the appropriate tone and language to use in text.
- Read the activity below, then answer the question accordingly.

Re-write the below paragraph to make it more appropriate for its audience. Think specifically about the language, tone and purpose of the text.

For the manager of RP Electrical.

Your electrician came round to fix my lights two weeks ago and they still don't work proper. The landing light wont switch off from the bottom of the stairs and he said he would be back the next day with a switch but he didn't. He also put mud on my carpet. It's not good enuff and I want some answers and my light fixed.

From

Mrs S V Johnson

Unit 6: Writing Skills

Section A: Formal letter writing

Short-answer questions

Instructions to students

- This exercise will help you to construct formal letters.
- Read the following example and then answer accordingly.

A formal letter is a method of communication that reflects how you or a business communicates in a formal manner. There are a number of functions and purposes and a letter of this nature should be clear, concise and courteous as well as following a structure. This can be seen below.

Key to Parts of a Formal Letter

a) The sender's address or, if sent by an organization, the letter heading of the company including a company logo.

b) Name, title and company name and address of the person and, if to a company, the company receiving the letter.

c) Date expressed as day, month and year.

d) Heading: indicating what the letter is about.

e) Salutation - Dear Mr/Mrs, etc. as the letter is addressed in the name and address line.

f) Introductory paragraph.

g) Middle paragraphs providing details.

h) Closing paragraphs providing an action statement and a courteous close.

i) Complimentary close: Yours sincerely because the recipient's name is used in the salutation. The writer's name and title, leaving space for the writer's signature.

127 Broadoak Drive, (a)
Clarendon Park,
Salisbury,
Wiltshire.
SP6 9ZZ

Head of Apprentice Recruitment, (b)
Human Resources Department,
Fast Response Electrical Engineering Ltd,
26 Winston Drive,
Andover,
Hampshire,
SP10 8XY

15 January 2012 (c)

Re: Apprenticeship application (d)

Dear Sir or Madam, (e)

I am writing to you in the hope that your company might be able to offer me an electrical apprenticeship, having seen an article in the local newspaper about your recent recruitment drive. (f)

The newspaper reported that your company was about to make an investment in young engineers and I would, very much, like to be a part of that programme if you would be kind enough to allow me the opportunity. (g)

I am 18 years old and I have had some work experience in electrical installation whilst I was at school. I really enjoyed it and learnt a lot of things about electricity. I worked hard at school and gained five A to C GCSEs including Mathematics (A), English Language (C) and Science (B). Since leaving school I have also achieved a Level 2 PEO in engineering at college. I have recently passed my driving test. Away from college I play for my local football team and I enjoy rock climbing and skiing. (g)

I would be very grateful if you were able to consider me for an apprenticeship, and would be very pleased to complete an application form, if you are indeed recruiting at this time or in the near future.

Thank you in advance for taking the time to read my letter. (h)

Yours sincerely (i)

Sign

PRINT YOUR NAME

Using the template, write a formal letter to a customer who has asked for a quote for work and materials for a replacement fuse box at her home. Briefly explain the work involved, the time it will take to complete the job and the price of the work. Address the letter to the customer, Mrs Burns of 8 Collins Street, Derby, DE1 8BQ.

Section B: Job application letter

Trainee Electrician, National Electrical Contractor

Location: Salisbury, Wiltshire

Salary: £14,000 – rising to £17,000 per annum on qualification

We are currently recruiting for a Trainee Electrician who has commitment, ambition and a real desire to learn about a full range of electrical installations. The role will be project based, working with experienced trades people on all aspects of commercial and electrical installations.

The successful applicant must have:

- A quality focus, taking pride in excellent workmanship

- Enthusiasm to work hard

- The ability to work as part of a team.

Please send your application to Mr Andrew Briars, Briars Electrical, North Way, Salisbury, Wiltshire

---

You have seen the above advert in your local paper with a vacancy for a trainee. Write a letter of application, setting out why you would like the job and the skills that you have that make you suitable for the job. Continue writing your letter using the notes section at the back of this workbook if required.

Section C: Persuasive writing

You are placing an advert on an auction website to sell your almost brand new electrical tool box. Write a full and honest description, highlighting all of the good points that will persuade someone to bid on your item. Think carefully about:

- the tone and language you use to make your item sound appealing to potential customers.

- making sure your description is clear and concise.

- the use of fact and opinion in your description.

Unit 7: Industry Writing Skills

Section A: Writing emails

Write an email to your customer, Mr Sandy Frey, reminding him that his combi boiler is due for its annual service in 30 days time.

To:
Subject:
Message:

Section B: Completing a risk assessment form

Fill in all of the sections of this risk assessment for a metal trunking installation on the 10th floor of an empty office block. The trunking run is to be through the centre of the office in the ceiling void and is to be installed before a suspended ceiling is erected.

RISK ASSESSMENT					NO.
ASSESSED BY:	DATE:				
AUTHORISED BY:	DATE:				
WORK ACTIVITY:					
TASK	HAZARD	LIKELY HARM	RISK RATING	CONTROL MEASURES REQUIRED	ADDITIONAL REQUIREMENTS

MATHEMATICS

It is important to show your workings out to indicate how you calculated your answer. Use this workbook to practice the questions and record your answers. Use extra paper if necessary to record your workings out.

Unit 8: General Mathematics

Short-answer questions

Instructions to students

- This unit will help you to improve your general mathematical skills.
- Read the following questions and answer all of them in the spaces provided.
- No calculators.
- You will need to show all working.

QUESTION 1

What unit of measurement would you use to measure:

a Electrical potential?

Answer:

_____

b The resistance in an electrical circuit?

Answer:

_____

c A length of conduit?

Answer:

_____

d The weight of a ladder?

Answer:

_____

e The speed of a broadband connection?

Answer:

_____

f The frequency of a current?

Answer:

_____

g The unit of electrical current?

Answer:

_____

QUESTION 2

Write an example of the following and give an example of where it may be found in the electrical industry.

a Percentages

Answer:

_____

b Decimals

Answer:

_____

c Fractions

Answer:

d Mixed numbers

Answer:

e Ratios

Answer:

f Angles

Answer:

QUESTION 3

Convert the following units:

a 12 kg to grams

Answer:

b 4 tonnes to kilograms

Answer:

c 1200 mm to metres

Answer:

d 1140 mL to litres

Answer:

e 1650 g to kilograms

Answer:

f 1880 kg to tonnes

Answer:

g 13 m to millimeters

Answer:

h 4.5 L to millilitres

Answer:

QUESTION 4

Write the following in descending order:

0.4   0.04   4.1   40.0   400.00   4.0

Answer:

QUESTION 5

Write the decimal number that is between:

a 0.2 and 0.4

Answer:

b 1.8 and 1.9

Answer:

c 12.4 and 12.6

Answer:

d 28.3 and 28.4

Answer:

e 101.5 and 101.7

Answer:

QUESTION 6

Round off the following numbers to two (2) decimal places.

a 12.346

Answer:

_____

b 2.251

Answer:

_____

c 123.897

Answer:

_____

d 688.882

Answer:

_____

e 1209.741

Answer:

_____

QUESTION 7

Estimate the following by approximation.

a $1288 \times 19 =$

Answer:

_____

b $201 \times 20 =$

Answer:

_____

c $497 \times 12.2 =$

Answer:

_____

d $1008 \times 10.3 =$

Answer:

_____

e $399 \times 22 =$

Answer:

_____

f $201 - 19 =$

Answer:

_____

g $502 - 61 =$

Answer:

_____

h $1003 - 49 =$

Answer:

_____

i $10\,001 - 199 =$

Answer:

_____

j $99.99 - 39.8 =$

Answer:

_____

QUESTION 8

What do the following add up to?

a £4, £4.99 and £144.95

Answer:

_____

b 8.75, 6.9 and 12.55

Answer:

_____

c 65 mL, 18 mL and 209 mL

Answer:

_____

d 21.3 g, 119 g and 884.65 g

Answer:

_____

QUESTION 9

Subtract the following.

a 2338 from 7117

Answer:

_____

b 1786 from 3112

Answer:

_____

c 5979 from 8014

Answer:

_____

d 11 989 from 26 221

Answer:

_____

e 108 767 from 231 111

Answer:

_____

QUESTION 10

Use division to solve:

a $2177 \div 7 =$

Answer:

_____

b $4484 \div 4 =$

Answer:

_____

c $63.9 \div 0.3 =$

Answer:

_____

d $121.63 \div 1.2 =$

Answer:

_____

e $466.88 \div 0.8 =$

Answer:

_____

The following information is provided for Question 11.

_____

To solve using BODMAS, in order from left to right, solve the Brackets first, then Orders, (Powers and Roots), then Division, then Multiplication, then Addition and lastly Subtraction. The following example has been done for your reference.

EXAMPLE

Solve $(4 \times 7) \times 2 + 6 - 4$.

_____

STEP 1

Solve the Brackets first: $(4 \times 7) = 28$

STEP 2

No Division, so next solve Multiplication: $28 \times 2 = 56$

STEP 3

Addition is next: $56 + 6 = 62$

STEP 4

Subtraction is the last process:

_____

FINAL ANSWER

58

_____

QUESTION 11

Using BODMAS, solve:

a $(6 \times 9) \times 5 + 7 - 2 =$

Answer:

_____

b $(9 \times 8) \times 4 + 6 - 1 =$

Answer:

___

c $3 \times (5 \times 7) + 11 - 8 =$

Answer:

___

d $6 + 9 - 5 \times (8 \times 3) =$

Answer:

___

e $9 - 7 + 6 \times 3 + (9 \times 6) =$

Answer:

___

f $(4 \times 3) - 6 + 9 \times 4 + (6 \times 7) =$

Answer:

___

g $(4 \times 9) - (3 \times 7) + 16 - 11 \times 2 =$

Answer:

___

h $9 - 4 \times 6 + (6 \times 7) + (8 \times 9) - 23 =$

Answer:

___

Unit 9: Basic Operations

Section A: Addition

Short-answer questions

Instructions to students

- This section will help you to improve your addition skills for basic operations.
- Read the following questions and answer all of them in the spaces provided.
- No calculators.
- You will need to show all working.

QUESTION 1

To rewire a lighting circuit an electrician uses
2 m, 1 m, 3 m and 5 m of electrical wire. How much wire
will be used in total?

Answer:

QUESTION 2

To rewire an office power circuit, an electrician uses 2.5
m, 1.8 m, 3.3 m and 15.2 m of electrical wire. How much
electrical wire will be used in total?

Answer:

QUESTION 3

An electrical shop stocks 127 of 15 W lamps, 368 of 10 W
lamps and 723 various lamps. How many lamps do they
have in stock in total?

Answer:

QUESTION 4

An electrician's van is driven 352 miles, 459 miles, 4872
miles and then 198 miles. How far has the van been
driven in total?

Answer:

QUESTION 5

An electrician uses the following amounts of diesel over
a month:

Week 1: 35.5 L

Week 2: 42.9 L

Week 3: 86.9 L

Week 4: 66.2 L

a How many litres have been used in total?

Answer:

b If diesel costs £1.95 per litre, how much would fuel
   have cost for the month?

Answer:

QUESTION 6

If an apprentice buys a crimping tool for £22.50,
four screwdrivers for £46.80 and a pair of pliers for
£6.75, how much has been spent?

Answer:

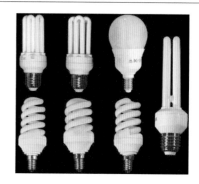

QUESTION 7

An electrician uses 10 mm nuts to complete three jobs. If he uses 26 on the first job, 52 on the second job and 48 on the third job, how many nuts have been used in total?

Answer:

QUESTION 8

An electrician buys a new multi-function test meter for £810.60, a cordless drill for £129.99 and a hole saw kit for £79.80. How much has been spent?

Answer:

QUESTION 9

An apprentice electrician travels 36.8 miles, 98.7 miles, 77.2 miles and 104.3 miles over 4 days. What distance has he covered in total?

Answer:

QUESTION 10

An electrician uses 178 bolts, 178 nuts and 356 washers to complete some mechanical work on a set of power boards. How many parts in total are needed?

Answer:

Section B: Subtraction

Short-answer questions

Instructions to students

- This section will help you to improve your subtraction skills for basic operations.
- Read the following questions and answer all of them in the spaces provided.
- No calculators.
- You will need to show all working.

QUESTION 1

A work van is filled up to its limit of 52 L with diesel. If the driver uses 22 L on one trip, 17 L on the second trip and 11 L on the third trip, how much is left in the tank?

Answer:

QUESTION 2

If an apprentice drives 362 miles and another driver covers 169 miles, how much further has the first driver gone?

Answer:

QUESTION 3

Driver A uses 243.8 L of diesel in a month, and Driver B uses 147.9 L of diesel in the same month. How much more does driver A use?

Answer:

QUESTION 4

An electrician uses 39 fuses from a box that originally contained 163 fuses. How many fuses are left?

Answer:

A work van service costs £224.65. The mechanic takes off a discount of approximately 10%, which he then rounds off to £25.00. What does the final bill come to after the discount?

Answer:

_____

QUESTION 6

Over a year, an apprentice drives 12,316 miles. Of this, 5787 miles is for his own personal use. How many miles did he travel for work-related purposes?

Answer:

_____

QUESTION 7

An electrician uses the following amounts of cable for three jobs:

Job 1: 5.5 m

Job 2: 3.8 m

Job 3: 6.9 m

If the drum of cable contained 50 m to begin with, how much cable is now left?

Answer:

_____

QUESTION 8

An electrician, Gary, replaces 74 fluorescent tubes in an office complex, during routine maintenance, in a month. If there were a total of 132 fluorescent tubes originally, how many fluorescent tubes are now left?

Answer:

_____

QUESTION 9

A car odometer has a reading of 78,769 miles. An apprentice then drives some distance to work on a bungalow's lighting circuit. When she returns the car, the odometer now reads 84,231 miles. How many miles have been travelled for the job?

Answer:

_____

QUESTION 10

An electrical apprentice uses the following amounts of the same type of cable on three separate jobs: 8.7 m, 6.9 m and 15.3 m. If there were 150 m of cable to begin with, how much cable would be left?

Answer:

_____

Section C: Multiplication

Short-answer questions

Instructions to students

- This section will help you to improve your multiplication skills for basic operations.
- Read the following questions and answer all of them in the spaces provided.
- No calculators.
- You will need to show all working.

QUESTION 1

If a car travels at 60 miles/h, how far will it travel in 9 hours?

Answer:

_____

QUESTION 2

If a car travels at 80 miles/hr, how far will it travel in 3 hours?

Answer:

_____

QUESTION 3

An electrician uses 15 L of fuel for a trip to a job. How much fuel will he use if the same trip needs to be completed 26 times?

Answer:

_____

QUESTION 4

An electrician uses four nuts, eight washers and four bolts to secure one (1) circuit-breaker. How many nuts, washers and bolts would be used on 144 circuit-breakers?

Answer:

_____

QUESTION 5

If 1.5 m of green/yellow wire, 2.2 m of brown wire and 0.8 m of blue wire are used to install a two-way lighting circuit in an office, how much of each wire would be used for 12 similar offices?

Answer:

_____

_____

QUESTION 6

If 24 wheel nuts are used to secure four wheels on to one (1) work van, how many nuts would you need for eight vans?

Answer:

_____

QUESTION 7

An electrician's truck uses 9 L of diesel for every 100 km she travels. How much diesel would she use to travel 450 km?

Answer:

_____

QUESTION 8

The assembly line of a company making kettles installs 673 fuses per month. If the same amount were installed each month, how many fuses would be installed over a year?

Answer:

_____

QUESTION 9

If an electrician installs an average of 32 m of cable each day over 28 days, how much cable has been used in total?

Answer:

_____

QUESTION 10

If a car travels at 110 miles/hr for 5 hours, how far has it travelled in total?

Answer:

_____

Section D: Division

Short-answer questions

Instructions to students

- This section will help you to improve your division skills for basic operations.
- Read the following questions and answer all of them in the spaces provided.
- No calculators.
- You will need to show all working.

QUESTION 1

An electrician has 24 m of cable. How many jobs can be completed if each standard job requires 3 m of cable?

Answer:

QUESTION 2

If an electrician earns £868 for working a 5-day week, how much does she earn per day?

Answer:

QUESTION 3

An electrical company buys 1400 m of cable in bulk. Each roll of cable contains 50 m.

How many full rolls are there?

Answer:

QUESTION 4

A contract electrician covers 780 miles in a 5-day week. On average, how many miles per day have been travelled?

Answer:

QUESTION 5

The total weight of a work van is 2488 kg. How much load, in kilograms, is on each of the four wheels?

Answer:

QUESTION 6

An electrical contractor covers 1925 miles over a 7-day period. How many miles are covered, on average, each day?

Answer:

QUESTION 7

At a yearly stocktake, an apprentice electrician counts 648 conduit adaptors. If there are 12 conduit adaptors in each box, how many boxes are there?

Answer:

QUESTION 8

An electrician orders 408 lamps. If there are 24 lamps in each box, how many boxes are there?

Answer:

QUESTION 9

An assembly line produces 680 power boards. The power boards will be used at 34 different locations. How many will be allocated to each location?

Answer:

QUESTION 10

An apprentice has 560 resistors. Of these, 28 are used on one job. How many identical jobs can be completed in total?

Answer:

Unit 10: Decimals

Section A: Addition

Short-answer questions

Instructions to students

- This section will help you to improve your addition skills when working with decimals.
- Read the following questions and answer all of them in the spaces provided.
- No calculators.
- You will need to show all working.

QUESTION 1

A set of four capacitors are purchased for £46.88 and a pair of pliers for £4.75. How much will be paid in total?

Answer:

QUESTION 2

An electrician buys four batten light fittings for £2.49 each, a 50 m roll of cable for £13.50, a box of two-way plate switches for £9.00 and a box of one-gang pattresses for £7.00. How has he spent in total?

Answer:

QUESTION 3

One length of conduit measures 9.85 m. Another length measures 12.75 m. If they need to be joined together, what will be the total length?

Answer:

QUESTION 4

Two lengths of 25 mm conduit are to be joined together. The first length measures 1025 mm while the second measures 848 mm.

a What is the total length in metres?

Answer:

b If they were both cut from the one original length measuring 2 m, how much is left over?

Answer:

QUESTION 5

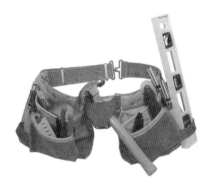

An electrician buys the following: a tool belt for £38.99, a pair of pliers for £8.95, a spirit level for £6.99 and a claw hammer for £10.60. What is the total cost?

Answer:

QUESTION 6

If a contract electrician travels 65.8 miles, 36.5 miles, 22.7 miles and 89.9 miles over 4 days, how far has he travelled in total?

Answer:

QUESTION 7

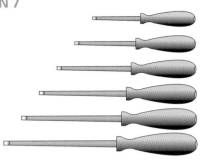

What is the total length of a screwdriver with an insulated handle of 115 mm and a shaft of 78 mm?

Answer:

QUESTION 8

An electrician has an assortment of conduit fittings in the back of his van. There are 22 conduit adaptors, 12 'T' boxes, four end boxes, seven inspection elbows and 35 couplers. How many conduit fittings are there in total?

Answer:

QUESTION 9

An electrician completes three jobs. He charges £450.80 for the first job, £1130.65 for the second job and £660.45 for the third job. How much has he charged in total for all three jobs?

Answer:

QUESTION 10

A room is 4.545 m long and 3.250 m wide. The electrician is required to install a trunking system around the parameter of the room. What is the total length of the trunking system?

Answer:

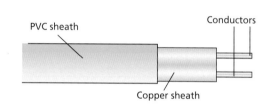

PVC sheath · Conductors · Copper sheath

Section B: Subtraction

Short-answer questions

Instructions to students

- This section will help you to improve your subtraction skills when working with decimals.
- Read the following questions and answer all of them in the spaces provided.
- No calculators.
- You will need to show all working.

QUESTION 1

An electrician trims a wire on a board. If the wire measures 380 mm and 110 mm is cut off, what length will be left?

Answer:

QUESTION 2

If an electrician cuts off 225 mm from a cable that is 1.5 m long, how much is left?

Answer:

QUESTION 3

An apprentice electrician completes a job that costs £789.20 and then a discount of £75.50 is given. How much is the final cost?

Answer:

_____

QUESTION 4

An apprentice 'sparky' works 38 hours and earns £245.60. On pay day he spends a total of £48.85 on his mobile phone bill, food and drink. How much is left?

Answer:

_____

QUESTION 5

A fan motor is lifted into place using a chain-block. The motor is lifted to 8.750 m and then lowered by 415 mm on to it's mounting plate. What height is the motor now at?

Answer:

_____

QUESTION 6

If a length of conduit that electrical wire is passed through has a diameter of 16 mm and a different length has a diameter of 30 mm, what is the difference between the two diameters?

Answer:

_____

QUESTION 7

The distance between each of four resistors in a series is 32.50 mm. The distance between two other resistors is 31.85 mm. What is the difference in distance?

Answer:

_____

QUESTION 8

An electrician has a 6 m length of SWA cable. He uses the length for three different jobs: 2.85 m for the first, 0.56 m for the next and 1.3 m for the last.

a How much is needed for all three jobs?

Answer:

_____

b How much is left over?

Answer:

_____

QUESTION 9

An electrician has two 10 m lengths of cable. He uses 3500 mm on one job, 7650 mm on another job, and then a further 4450 mm on the final job.

a How much cable is used overall?

Answer:

_____

b How much cable is left?

Answer:

_____

QUESTION 10

An apprentice electrician is making up pendant light sets. Each pendant set requires 370 mm of two-core flex. If the apprentice starts with 15 m of flex and makes 18 sets, how much flex is left?

Answer:

_____

Section C: Multiplication

Short-answer questions

Instructions to students

- This section will help you to improve your multiplication skills when working with decimals.
- Read the following questions and answer all of them in the spaces provided.
- No calculators.
- You will need to show all working.

QUESTION 1

If one tyre on an electrician's van costs £99.95 and he needs to purchase a total of five tyres (four for each wheel and one for the spare), how much will the total cost be?

Answer:

QUESTION 2

If an electrician uses 1 m of cable that costs £7.80 per metre, what is the cost of 15 m?

Answer:

QUESTION 3

An apprentice electrician replaces six fluorescent lights at a cost of £4.69 each. She then replaces eight circuit-breakers at a cost of £3.70 each. What is the total cost of the lights and the fuses?

Answer:

QUESTION 4

If an apprentice purchases six packets of 10 mm nuts that cost £8.65 per packet, how much is the total cost?

Answer:

QUESTION 5

An electrician buys 12 pairs of rigger's gloves that cost £5.39 each. How much did he spend in total?

Answer:

QUESTION 6

An electrician's hourly rate is £22.75. How much will she earn for a 45-hour week?

Answer:

QUESTION 7

The wholesaler has PVC conduit on special offer at £0.93 per 3 m length. An electrician buys 70 lengths. How much has he spent on PVC conduit?

Answer:

QUESTION 8

A contractor fills up their 52 L car petrol tank at £1.55 per litre. How much does the contractor pay for the petrol?

Answer:

QUESTION 9

An electrical company purchases 340 two-gang surface boxes. The cost of each box is £1.23 What is the outlay?

Answer:

QUESTION 10

An apprentice electrician earns £80.65 per day. How much is his gross pay for a 5-day week?

Answer:

Section D: Division

Short-answer questions

Instructions to students

- This section will help you to improve your division skills when working with decimals.
- Read the following questions and answer all of them in the spaces provided.
- No calculators.
- You will need to show all working.

QUESTION 1

An electrician is to wire-up six new houses. He has a total of 288 two-gang switched socket-outlets to install. If the socket-outlets are evenly divided between the six houses, how many are installed in each house?

Answer:

QUESTION 2

An electrician earns £1590.60 for 5 days of work. How much does she earn per day?

Answer:

QUESTION 3

An electrical contractor travels 525 miles over 3 days. How far does he travel, on average, each day?

Answer:

QUESTION 4

An electrician completes a job worth £440.85. If the job takes 16 hours to complete, what is the hourly rate?

Answer:

QUESTION 5

A truck driver drives from Dover to Carlisle via Bournemouth with electrical supplies. He covers 536.84 miles over 2 days. If his total driving time is 15 hours, how far has he travelled, on average, each hour?

Answer:

QUESTION 6

An apprentice electrician drives from Exeter to King's Lynn for a training course. She travels a total of 278.85 miles over 5 hours. What distance did she travel, on average, each hour?

Answer:

QUESTION 7

A work vehicle uses 36 L to travel 288.8 miles. How far does the car travel per litre?

Answer:

QUESTION 8

An electrician orders 360 fluorescent luminaires at a cost of £1890 for a workshop. How much is the cost of one fluorescent luminaire?

Answer:

QUESTION 9

It costs £80.95 to fill a work van's 52 L fuel tank. How much is the cost per litre?

Answer:

QUESTION 10

A 50 m roll of cable costs £83.60. How much does the cable cost per metre?

Answer:

Unit 11: Fractions

Section A: Addition

Short-answer questions

Specific instructions to students

- This section is designed to help you to improve your addition skills when working with fractions.
- Read the following questions and answer all of them in the spaces provided.
- No calculators.
- You will need to show all working.

QUESTION 1

$\frac{1}{2} + \frac{4}{5} =$

Answer:

QUESTION 2

$2\frac{2}{4} + 1\frac{2}{3} =$

Answer:

QUESTION 3

A bolt is inserted $\frac{2}{4}$ of the way through a wall. If it is pushed another $\frac{1}{5}$ through, how far has the bolt been pushed through the wall? Express your answer as a fraction.

Answer:

QUESTION 4

A gang nail is hammered $\frac{1}{3}$ of the way into a gyprock wall. It is then hammered a further $\frac{2}{5}$ into the wall. How far has it gone into the wall? Express your answer as a fraction.

Answer:

QUESTION 5

An aerial connecting wire is fed $\frac{3}{4}$ of the way through a hole in a wall. If it is then fed a further $\frac{1}{6}$ in, how far has the aerial been fed through the wall in total? Express your answer as a fraction.

Answer:

Section B: Subtraction

Short-answer questions

Instructions to students

- This section is designed to help you to improve your subtraction skills when working with fractions.
- Read the following questions and answer all of them in the spaces provided.
- No calculators.
- You will need to show all working.

QUESTION 1

$\frac{2}{3} - \frac{1}{4} =$

Answer:

_____

QUESTION 2

$2\frac{2}{3} - 1\frac{1}{4} =$

Answer:

_____

QUESTION 3

An electrician has $2\frac{2}{3}$ rolls of cable. $1\frac{1}{2}$ are used on a housing job. How much cable now remains? Express your answer as a fraction.

Answer:

_____

QUESTION 4

An apprentice has $5\frac{1}{2}$ packets of screws. If $3\frac{1}{3}$ packets are used on a repair job, how many boxes are left? Express your answer as a fraction.

Answer:

_____

QUESTION 5

An electrician takes $2\frac{1}{2}$ rolls of cable to an industrial complex for use on a repair job. The electrician uses $1\frac{1}{3}$ of the rolls to complete the job. How much cable is left? Express your answer as a fraction.

Answer:

_____

Section C: Multiplication

Short-answer questions

Instructions to students

- This section is designed to help you improve your multiplication skills when working with fractions.
- Read the following questions and answer all of them in the spaces provided.
- No calculators.
- You will need to show all working.

QUESTION 1

$\frac{2}{4} \times \frac{2}{3} =$

Answer:

QUESTION 2

$2\frac{2}{3} \times 1\frac{1}{2} =$

Answer:

QUESTION 3

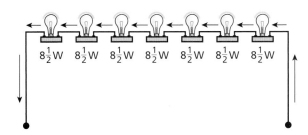

$8\frac{1}{2}$ W $8\frac{1}{2}$ W $8\frac{1}{2}$ W $8\frac{1}{2}$ W $8\frac{1}{2}$ W $8\frac{1}{2}$ W $8\frac{1}{2}$ W

Seven lights are connected in series. If each light uses $8\frac{1}{2}$ W, how many watts are there in total? Express your answer as a fraction.

Answer:

_____

QUESTION 4

A battery weighs $1\frac{1}{2}$ kg. How much will 12 batteries weigh?

Answer:

_____

QUESTION 5

A wiring job requires 6 lengths of brown wire that each measure $18\frac{1}{2}$ m, 8 lengths of green/yellow wire that each measure $16\frac{1}{2}$ m, and 4 lengths of blue wire that each measure $20\frac{1}{2}$ m. How much of the brown, green/yellow and blue wire is needed to complete the job?

Answer:

_____

_____

Section D: Division

Short-answer questions

Instructions to students

- This section is designed to help you to improve your division skills when working with fractions.
- Read the following questions and answer all of them in the spaces provided.
- No calculators.
- You will need to show all working.

QUESTION 1

$$\frac{2}{3} \div \frac{1}{4} =$$

Answer:

_____

QUESTION 2

$$2\frac{3}{4} \div 1\frac{1}{3} =$$

Answer:

_____

QUESTION 3

An apprentice has a 50 m roll of electrical cable. It needs to be cut into $6\frac{1}{2}$ m lengths.

a How many lengths can be cut?

Answer:

_____

b Is there any wastage?

Answer:

_____

QUESTION 4

An electrician needs to cut $3\frac{1}{2}$ cm machine bolt blanks from a 2 m length of steel stock. How many bolt blanks can be cut?

Answer:

_____

QUESTION 5

An electrician has $6\frac{1}{2}$ packets of screws that are to be used on four separate jobs. How many packets will be needed for each job? Express your answer as a fraction.

Answer:

_____

Unit 12: Percentages

Short-answer questions

Instructions to students

- In this unit, you will be able to practise and improve your skills in working out percentages.
- Read the following questions and answer all of them in the spaces provided.
- No calculators.
- You will need to show all working.

> 10% rule: Move the decimal one place to the left to get 10%.
>
> **Example**
>
> 10% of £45.00 would be £4.50

QUESTION 1

An electrical repair bill comes to £1220.00. How much is 10% of the bill?

Answer:

_____

QUESTION 2

A CCTV camera is installed at a cost of £549.00 for parts and labour.

a What is 10% of the cost?

Answer:

_____

b If this 10% is a discount, what will the final cost be?

Answer:

_____

QUESTION 3

The owner of an electrical workshop buys a 2 hp direct drive air compressor for £198.50.

a If he was given a 10% discount, how much would the discount work out to be?

Answer:

_____

b What would the cost of the air compressor be after the discount?

Answer:

_____

QUESTION 4

An electrician buys five rolls of cable for a total of £124.80. She is given a 5% discount by the shop owner. How much will the electrician need to pay after the discount? (Hint: Find 10%, halve it and then subtract it from £124.80.)

Answer:

_____

QUESTION 5

An installer buys three storage bins for £42, an 18 V drill for £280 and a set of allen keys for £16.

a How much is the total?

Answer:

_____

b How much is paid after a 10% discount?

Answer:

_____

QUESTION 6

The following items are purchased for an additional ring final circuit installation: 50m of 2.5 mm² twin and earth cable, for £24.85, a 32A RCBO for £23.30, two one-gang moulded boxes, £1.09 each, eight two-gang moulded boxes at £1.85 each, two 13A switch, fused connection units, at £5.05 each, eight double socket outlets at £2.56 each and six 3m lengths of mini-trunking at £1.68 each.

a What is the total cost of all of the items?

Answer:

b What is the final cost after a 10% discount?

Answer:

QUESTION 7

An electrical store offers 20% off the price of sets of screwdrivers. If a set is priced at £136 before the discount, how much will they cost after the discount?

Answer:

QUESTION 8

Approved voltage testers are discounted by 15%. If the regular retail price is £36 each, what is the discounted price?

Answer:

QUESTION 9

The regular retail price of a set of industrial drill bits costs £186.80. The store then has a 20% off sale. How much will the drill bits cost during the sale?

Answer:

QUESTION 10

A power drill costs £99 at the wholesaler. How much will it cost after the store takes off 30%?

Answer:

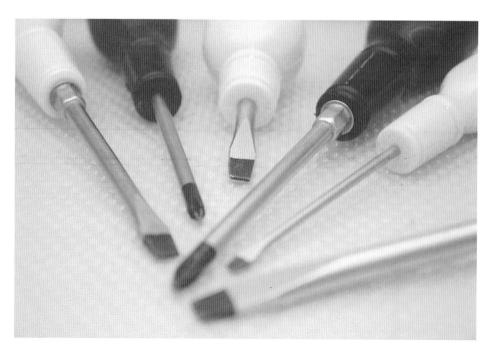

Unit 13: Voltage, Current and Resistance

Short-answer questions

Instructions to students

- In this unit, you will be able to practise and improve your skills in calculating voltage, current and resistance.
- Read the following questions and answer all of them in the spaces provided.
- No calculators.
- You will need to show all working.

Ohm's law: $V = I \times R$

where:

V = voltage, with volts (V) as the unit of measurement

I = current, with amperes (A) as the unit of measurement

R = resistance, with ohms (Ω) as the unit of measurement

Transposing this formula gives the following variations:

$I = \dfrac{V}{R}$

$R = \dfrac{V}{I}$

QUESTION 1

What is the voltage (V) of a small appliance if the current (I) is 12 A and the resistance (R) is 10 Ω?

Answer:

QUESTION 2

What is the resistance (R) if the current (I) is 15 A and the voltage (V) is 230 V?

Answer:

QUESTION 3

Find the current (I) if the voltage (V) is 230 V and the resistance (R) is 20 Ω.

Answer:

QUESTION 4

Find the voltage (V) if the resistance (R) is 25 Ω and the current (I) is 5 A.

Answer:

QUESTION 5

What is the resistance (R) if the current (I) is 25 A and the voltage (V) is 230 V?

Answer:

QUESTION 6

Find the current (I) if the voltage (V) is 230 V and the resistance (R) is 50 Ω.

Answer:

QUESTION 7

Find the voltage (V) if the resistance (R) is 35 Ω and the current (I) is 4 A.

Answer:

QUESTION 8

What is the resistance (R) if the current (I) is 15 A and the voltage (V) is 24 V?

Answer:

QUESTION 9

Find the current (I) if the voltage (V) is 12 V and the resistance (R) is 0.5 Ω.

Answer:

QUESTION 10

Find the voltage (V) if the resistance (R) is 1.5 Ω and the current (I) is 3 A.

Answer:

For Questions 11–15, note that the total resistance equals the sum of the resistors in series.

QUESTION 11

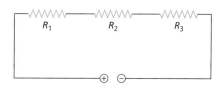

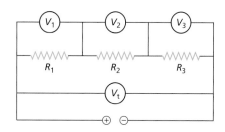

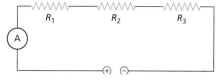

Find the total resistance (R_t) if resistor 1 (R_1) equals 130 Ω, resistor 2 (R_2) equals 100 Ω and resistor 3 (R_3) equals 180 Ω.

Answer:

QUESTION 12

Find the total resistance (R_t) if R_1 equals 60 Ω, R_2 equals 110 Ω and R_3 equals 100 Ω.

Answer:

QUESTION 13

Find the total resistance (R_t) if R_1 equals 0.03 Ω, R_2 equals 1.70 Ω and R_3 equals 11.0 Ω.

Answer:

QUESTION 14

Find the total resistance (R_t) if R_1 equals 4.25 Ω, R_2 equals 7.5 Ω and R_3 equals 13.7 Ω.

Answer:

QUESTION 15

Find the total resistance (R_t) if R_1 equals 0.5 Ω, R_2 equals 2.5 Ω and R_3 equals 3.5 Ω.

Answer:

Unit 14: Measurement Conversions and SI Units

Section A: General measurement conversions

Short-answer questions

Instructions to students

- This unit is designed to help you to both improve your skills and increase your speed in converting one measurement unit into another.
- This unit is also designed to help you use multiples and sub-multiples of engineering units and identify standard engineering form associated with electrical engineering correctly.
- Read the following questions and answer all of them in the spaces provided.
- No calculators.
- You will need to show all working.

QUESTION 1

How many millimetres are there in 0.01 m?

Answer:

QUESTION 2

How many millimetres are there in 1 m?

Answer:

QUESTION 3

How many millimeters are there in 2 m?

Answer:

QUESTION 4

If a screw has 20 threads in 20 mm of its length, how many threads would there be in 100 mm?

Answer:

QUESTION 5

How many millilitres are there in 4.8 L of de-greaser?

Answer:

QUESTION 6

How many litres does 3500 mL of bearing grease make up?

Answer:

QUESTION 7

An electric motor weighs ½ a tonne. How many kilograms is that?

Answer:

QUESTION 8

A drum of underground power cable weighs 2 tonnes. How many kilograms is that?

Answer:

QUESTION 9

A truck weighs 4750 kg. How many tonnes is that?

Answer:

QUESTION 10

The wall of a shed that will have wiring running around its perimeter measures 1800 mm in length and 1200 mm in width. How far is it around the perimeter of the wall?

Answer:

Section B: Electrical engineering units and quantities

In 1968, British industry adopted the SI system of measurement (Systeme International d'unites), to replace the old imperial system and align with her European partners. This way a single standard could be applied throughout Europe.

The SI system uses the number 10, its multiples and sub-multiples as its numerical base. SI units are used to define time, length, mass, area, volume, speed, acceleration and force related to time.

BASE UNITS.

A base unit is a unit that has been defined in some way by man and is quantifiable in some way by a given standard. E.g. the base unit of length is the metre; the metre is defined as the distance travelled by a red light of a given frequency in a vacuum in a given time. The speed of light is give as 1/299792458 m/s exactly.

BASE UNITS			
Quantity	Quantity Symbol	Unit	Unit Symbol
Length	l	Metre	m
Mass	m	grams	g
Time	t	Seconds	s
Electric current	I	Ampere	A
Temperature	T	Kelvin	K

DERIVED UNITS.

Derived units are units are those formed by combining base quantities; the Coulomb (C) is the quantity of electricity transported in 1 second by a current of 1 Ampere.

1 Coulomb = 1 Ampere second

DERIVED UNITS			
Quantity	Quantity Symbol	Unit	Unit Symbol
Electric Capacitance	C	Farad	F
Electric Charge	Q	Coulomb	C
Electric Potential	e	Volt	V
Electric Resistance	R	Ohm	Ω
Energy	E	Joule	J
Frequency	F	Hertz	Hz
Power	P	Watt	W

Short-answer questions

Instructions to students

- This exercise will help you to identify SI units associated with electrical engineering.
- Read the following question and then answer accordingly.

Complete the following statements.

1. Mass (m) is measured in _____

2. _____ is measured in Hertz (Hz)

3. Electric current ___ is measured in Amperes (A)

4. Power (P) is measured in Watts ___

5. _____ is measured in Coulombs (C)

6. Capacitance (C) is measured in _____

7. A base unit is defined as a _____ standard

8. A derived unit is formed by _____

Section C: Standard engineering form

Short-answer questions

Instructions to students

- This exercise will help you to use multiples and sub-multiples of engineering units correctly.
- Read the following question and then answer accordingly.
- This section will help you to identify standard engineering form associated with electrical engineering correctly.

Multiple and sub-multiple units.

When we measure quantities in engineering, occasionally the result is convenient, e.g. 6 volts or 22 amps, however there are situations where the measurement is very small or very large and this is much less convenient to record, e.g. 100 000 volts or 0.000 003 ohms. To make it easier to write down very large or very small numbers, we use a system of multiple and sub-multiple prefixes.

For very large numbers we use multiples of 10.

For very small numbers we use sub-multiples of 10.

*note that in engineering we do not use; hecto, deca, deci or centi.

The table below has a list of multiples and sub-multiples with their related prefixes and symbols.

PREFIXES TO CREATE MULTIPLES AND SUB-MULTIPLES OF SI UNITS.			
MULTIPLIER	EXPONENT FORM	PREFIX	SI SYMBOL
1 000 000 000 000 000 000 000 000	10^{24}	yotta	Y
1 000 000 000 000 000 000 000	10^{21}	zetta	Z
1 000 000 000 000 000 000	10^{18}	exa	E
1 000 000 000 000 000	10^{15}	peta	P
1 000 000 000 000	10^{12}	tera	T
1 000 000 000	10^{9}	giga	G
1 000 000	10^{6}	mega	M
1 000	10^{3}	kilo	k
100 NOT USED IN ENGINEERING*	10^{2}	hecto	h
10 NOT USED IN ENGINEERING	10^{1}	deca	da
1	10^{0}	unity	
0×1 NOT USED IN ENGINEERING	10^{-1}	deci	d
0×01 NOT USED IN ENGINEERING	10^{-2}	centi	c
0×001	10^{-3}	milli	m
0×000 001	10^{-6}	micro	m
0×000 000 001	10^{-9}	nano	n
0×000 000 000 001	10^{-12}	pico	P
0×000 000 000 000 001	10^{-15}	femto	f
0×000 000 000 000 000 001	10^{-18}	atto	a
0×000 000 000 000 000 000 001	10^{-21}	zepto	z
0×000 000 000 000 000 000 000 001	10^{-24}	yocto	y

Match the numbers to the correct prefix

0.000 000 001		kilo
0.000 001		giga
0.001		nano
1		mega
1000		micro
1000 000		unity
1000 000 000		milli

Complete the following sentences

1 There are _____ millimetres in 2 metres.

2 4 megawatts is 4000 000 _____

3 There are 1500 volts in 1.5 _____

4 3000 nanofarads is 3 _____

Unit 15: Measurement – Length and Area

Section A: Circumference

Short-answer questions

Instructions to students

- This section is designed to help you to both improve your skills and increase your speed in measuring the circumference of a round object.
- Read the following questions and answer all of them in the spaces provided.
- No calculators.
- You will need to show all working.

$C = \pi \times d$

where: C = circumference, π = 3.14 and d = diameter

EXAMPLE

Find the circumference of a plate with a diameter of 300 mm.

$C = \pi \times d$

Therefore, $C = 3.14 \times 300$

$\qquad = 942$ mm

QUESTION 1

Find the circumference of a warehouse light fitting with a diameter of 600 mm.

Answer:

QUESTION 2

Calculate the circumference of a pulley with a diameter of 150 mm.

Answer:

QUESTION 3

Determine the circumference of a store light fitting with a diameter of 320 mm.

Answer:

QUESTION 4

Find the circumference of a conduit with a diameter of 50 mm.

Answer:

QUESTION 5

Calculate the circumference of a heat pump pipe with a diameter of 120 mm.

Answer:

QUESTION 6

Determine the circumference of a spotlight fitting with a diameter of 288 mm.

Answer:

QUESTION 7

Find the circumference of a car speaker hole with a diameter of 156 mm.

Answer:

QUESTION 8

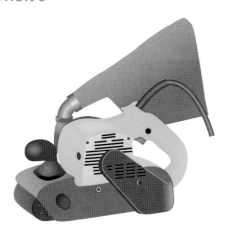

Determine the circumference of a 1200 W sander with a diameter of 143 mm.

Answer:

QUESTION 9

Find the circumference of an industrial pulley with a diameter of 429 mm.

Answer:

QUESTION 10

Calculate the circumference of a 50 W floodlight with a diameter of 188 mm.

Answer:

Section B: Diameter

Short-answer questions

Instructions to students

- This section is designed to help you to both improve your skills and increase your speed in measuring the diameter of a round object.
- Read the following questions and answer all of them in the spaces provided.
- No calculators.
- You will need to show all working.

Diameter (*D*) of a circle $= \dfrac{\text{circumference}}{\pi(3.14)}$

EXAMPLE

Find the diameter of a conduit with a circumference of 800 mm.

$$D = \frac{800}{3.14} = 254.77 \text{ mm}$$

QUESTION 1

Find the diameter of a conduit with a circumference of 220 mm.

Answer:

QUESTION 2

Calculate the diameter of a ceiling hole for a light with a circumference of 160 mm.

Answer:

QUESTION 3

Determine the diameter of a CCTV camera head with a circumference of 200 mm.

Answer:

QUESTION 4

Find the diameter of a cable drum with a circumference of 785 mm.

Answer:

QUESTION 5

Calculate the diameter of a floodlight with a circumference of 500 mm.

Answer:

QUESTION 6

Determine the diameter of a fuel tank with a circumference 11.8 m.

Answer:

QUESTION 7

Find the diameter of a pipe with a circumference of 1244 mm.

Answer:

QUESTION 8

Calculate the diameter of a warehouse light fitting with a circumference of 908 mm.

Answer:

QUESTION 9

Determine the diameter of a cog with a circumference of 623 mm.

Answer:

QUESTION 10

Find the diameter of a storeroom light fitting with a circumference of 688 mm.

Answer:

Section C: Area

Short-answer questions

Instructions to students

- This section is designed to help you to both improve your skills and increase your speed in measuring surface area.
- Read the following questions and answer all of them in the spaces provided.
- No calculators.
- You will need to show all working.

> **Area = length × breadth and is given in square units**
>
> **= *l* × *b***

QUESTION 1

An electrician needs to transport trunking in his trailer. The length of the electrician's trailer is 1.8 m by 1.2 m wide. What is the total floor area?

Answer:

QUESTION 2

If an electrical workshop measures 60 m by 13 m, what is the total area?

Answer:

QUESTION 3

A light switch cover is 128 mm by 128 mm. What is its total area?

Answer:

QUESTION 4

If a switchboard measures 4.5 m by 3.8 m, what is its total area?

Answer:

QUESTION 5

A fan control panel measures 120 mm by 100 mm, what is the total area?

Answer:

QUESTION 6

A battery has plates inside of it that measure 155 mm by 128 mm. What is the total area of one plate?

Answer:

QUESTION 7

The dimensions of the floor of an electrician's van are 1.06 m by 1.07 m. What is the total area?

Answer:

QUESTION 8

An electrical warehouse storage area is 65.3 m by 32.7 m. How much floor area is there?

Answer:

QUESTION 9

If the floor of a garage is 3.2 m wide by 8.6 m long, what is its area?

Answer:

QUESTION 10

An electrical spare parts delivery truck is 8.9 m long and 2.6 m wide. How much floor area can it accommodate?

Answer:

Unit 16: Earning Wages

Short-answer questions

Instructions to students

- This unit will help you to calculate how much a job is worth and how long you need to complete it.
- Read the following questions and answer all of them in the spaces provided.
- No calculators.
- You will need to show all working.

QUESTION 1

A first-year apprentice electrician may earn £270.45 clear per week. Based on this weekly salary, how much would the apprentice electrician earn per year? (Note that there are 52 weeks per year.)

Answer:

QUESTION 2

An apprentice 'sparky' starts work at 8.00 a.m. and stops for a break at 10.30 a.m. for ½ an hour. He goes back to work and steadily continues until 1.15 p.m. when he stops for a lunch break for ¾ of an hour. After lunch he works through to 4.00 p.m. How many hours has he worked?

Answer:

QUESTION 3

An electrician's mate earns £15.50 an hour and works a 38-hour week. How much are her gross earnings (before tax)?

Answer:

QUESTION 4

Over a week, an electrician completes five jobs, which are billed as: £465.80, £2490.50, £556.20, £1560.70 and £990.60. What do his total bills come to?

Answer:

QUESTION 5

An apprentice electrician needs to complete the following tasks:

- Task 1: to safely isolate, lock off and test a motor control circuit and erect the appropriate signs, which takes 34 minutes to complete.
- Task 2: to remove a set of wires, which takes 18 minutes.
- Task 3: to replace the contactors and thermal protection relays, which takes 27 minutes.
- Task 4: to replace and terminate the wiring, which takes 44 minutes.
- Task 5: to test the installation and reset the thermal overload relays, which takes 9 minutes.

How much time, in total, is needed to complete all of these tasks? State your answer in hours and minutes.

Answer:

QUESTION 6

The front room of a house needs to be rewired. This takes the electrician 4½ hours. If the pay rate is £38.50 an hour, how much will the electrician earn?

Answer:

QUESTION 7

Changing a broken socket outlet takes 1½ hours to complete. If the apprentice is getting paid £14.80 per hour, what amount will he earn for this job?

Answer:

QUESTION 8

A house has electrical damage due to a lightning strike. It takes the electrician 116 hours of work to return it to a safe and workable condition. If the electrician works 8-hour days, how many days did it take?

Answer:

QUESTION 9

An apprentice begins work at 7.00 a.m. and works until 3.30 p.m. The morning break is 20 minutes, the lunch break is 60 minutes and the afternoon break is 20 minutes.

a How much time has been spent on breaks?

Answer:

b How much time has been spent working?

Answer:

QUESTION 10

A major electrical job costs £2850.50 to complete. The apprentice spends 100 hours on the job. How much is the rate per hour?

Answer:

Unit 17: Squaring Numbers

Section A: Introducing square numbers

Short-answer questions

Instructions to students

- This section is designed to help you to both improve your skills and increase your speed in squaring numbers.
- Read the following questions and answer all of them in the spaces provided.
- No calculators.
- You will need to show all working.

> **Any number squared is multiplied by itself.**

EXAMPLE

4 squared $= 4^2 = 4 \times 4 = 16$

QUESTION 1

$6^2 =$

Answer:

QUESTION 2

$8^2 =$

Answer:

QUESTION 3

$12^2 =$

Answer:

QUESTION 4

$3^2 =$

Answer:

QUESTION 5

$7^2 =$

Answer:

QUESTION 6

$11^2 =$

Answer:

QUESTION 7

$10^2 =$

Answer:

QUESTION 8

$9^2 =$

Answer:

QUESTION 9

$2^2 =$

Answer:

$4^2 =$

Answer:

_____

$5^2 =$

Answer:

_____

Section B: Applying square numbers to the trade

Worded practical problems

Instructions to students

- This section is designed to help you to both improve your skills and increase your speed in calculating the area of rectangular or square objects.
- No calculators.
- You will need to show all working.

QUESTION 1

An apprentice sets aside a work area to work on a switchboard. The area measures 2.8 m × 2.8 m. What floor area does it take up?

Answer:

_____

QUESTION 2

A workshop has a control room area that is 5.2 m × 5.2 m. What is the total floor area?

Answer:

_____

QUESTION 3

The dimensions of an electrical workshop are 12.6 m × 12.6 m. What is the total floor area?

Answer:

_____

QUESTION 4

An electrician works in an area where some rewiring is needed. The floor area is 15 m × 15 m. If the area allocated for the storage of electrical tools measures 2.4 m × 2.4 m and does not require rewiring, how much area is left to rewire?

Answer:

_____

QUESTION 5

An electrician has a total work area of 13.8 m × 13.8 m. If the spare parts area takes up 1.2 m × 1.2 m and the tool area is 2.7 m × 2.7 m, how much area is left to work in?

Answer:

_____

QUESTION 6

An apprentice needs to construct a control panel from material that measures 2.4 m × 2.4 m. If 1.65 m × 1.65 m is cut out initially, how much is left?

Answer:

_____

QUESTION 7

An electrician cuts out a panel 0.5 m × 0.5 m from a sheet that is 1.2 m × 1.2 m. How much is left?

Answer:

_____

QUESTION 8

A concrete work floor of an electrical workshop measures 28.2 m × 28.2 m. If it costs £9.50 to coat 1 m², how much will it cost to coat the whole floor?

Answer:

_____

_____

QUESTION 9

Each wall of a welding area measures 2.6 m × 2.6 m. To insulate 1 m² it costs £28.50. How much will it cost to insulate all four walls?

Answer:

_____

_____

QUESTION 10

A distribution trunking needs to be installed around the perimeter of a room. The wall measures 3.2 m × 3.2 m.

a How much wire is required to go around the outside of three sides of the wall?

Answer:

_____

b If three different wires need to be installed in the trunking, what is the total amount of wire required?

Answer:

_____

Section A: Introducing ratios

Short-answer questions

Instructions to students

- This section is designed to help improve your skills in calculating and simplifying ratios.
- Read the following questions and answer all of them in the spaces provided.
- No calculators.
- You will need to show all working.
- Reduce the ratios to the simplest or lowest form.

QUESTION 1

The number of teeth on gear cog 1 is 40. The number of teeth on gear cog 2 is 20. What is the ratio of gear cog 1 to gear cog 2?

Answer:

QUESTION 2

Pulley A has a diameter of 600 mm and pulley B has a diameter of 150 mm. What is the ratio of diameter A to B?

Answer:

QUESTION 3

Pulley A has a diameter of 480 mm and pulley B has a diameter of 160 mm. What is the ratio of diameter A to B?

Answer:

QUESTION 4

A step-down transformer has 230 volts on winding A and 25.5 volts on winding B. What is the ratio of primary voltage to secondary voltage?

Answer:

QUESTION 5

Three cogs have 80 : 60 : 20 teeth respectively. What is the ratio?

Answer:

QUESTION 6

A lathe has two pulleys that have diameters of 160 mm and 200 mm respectively. What is the lowest ratio?

Answer:

QUESTION 7

The diameter of pulley A on a band saw is 320 mm. Pulley B has a diameter of 160 mm and pulley C has a diameter of 480 mm. What is the lowest ratio of the three compared together?

Answer:

QUESTION 8

Three pulleys have different diameters: 180 mm, 160 mm and 100 mm respectively. What is the comparative ratio?

Answer:

QUESTION 9

A step-down transformer has 680 turns on winding A and 240 turns on winding B. What is the turns ratio between the two windings?

Answer:

QUESTION 10

A current transformer has 10 amperes on winding A and 50 amperes on winding B. What is the ratio of primary current to secondary current?

Answer:

Section B: Applying ratios to the trade

Short-answer questions

Instructions to students

- This section is designed to help to improve your practical skills when working with ratios.
- Read the following questions and answer all of them in the spaces provided.
- No calculators.
- You will need to show all working.

QUESTION 1

The ratio of the teeth on cog 1 to cog 2 is 3 : 1. If cog 2 has 10 teeth, how many teeth will cog 1 have?

Answer:

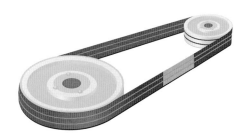

QUESTION 4

The ratio of the diameter of pulley A to pulley B is 2 : 1. If pulley A has a diameter of 300 mm, what will be the diameter of pulley B?

Answer:

QUESTION 2

The ratio of the teeth on cog 1 to cog 2 is 2 : 1. If cog 2 has 20 teeth, how many teeth will cog 1 have?

Answer:

QUESTION 5

The ratio of a transformer is 3:1. If there are 12 turns on winding A, how many turns will there be on winding B?

Answer:

QUESTION 3

The ratio of the diameter of pulley A to pulley B is 4 : 2. If pulley A has a diameter of 400 mm, what will be the diameter of pulley B?

Answer:

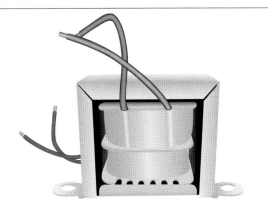

QUESTION 6

A transformer has a ratio of 2:1 if the are 220 turns on winding A, how many turns are there on winding B?

Answer:

QUESTION 7

A motor drive has a ratio of teeth on cog A to cog B of 3 : 1. If the number of teeth on cog A is 21, how many teeth will there be on cog B?

Answer:

QUESTION 8

The ratio of teeth on cog A to cog B is 3 : 2. If the number of teeth on cog A is 6, how many teeth will be on cog B?

Answer:

QUESTION 9

The ratio of windings in a transformer from winding A to winding B is 4:3. If winding A has 160 turns, how many turns are on winding B?

Answer:

QUESTION 10

The ratio of windings in a transformer from winding A to winding B is 7:2. If winding A has 252 turns, how many turns are on winding B?

Answer:

Unit 19: Mechanical Reasoning

Short-answer questions

Instructions to students

- This section is designed to help improve your skills in mechanical reasoning.
- Read the following questions and answer all of them in the spaces provided.
- No calculators.
- You will need to show all working.

QUESTION 1

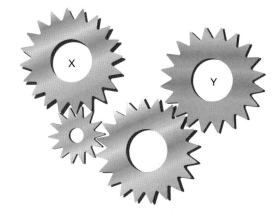

If cog X turns in a clockwise direction, which way will cog Y turn?

Answer:

_____

QUESTION 2

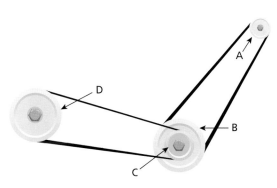

If pulley A turns in a clockwise direction, which way will pulley D turn?

Answer:

_____

QUESTION 3

If the drive pulley in the following diagram of a work van engine turns in a clockwise direction, in which direction will the alternator turn?

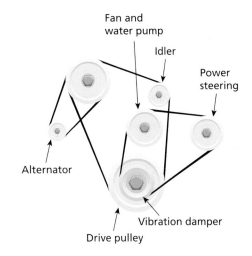

Answer:

_____

QUESTION 4

Looking at the following diagram, if lever A moves to the left, in which direction will lever B move?

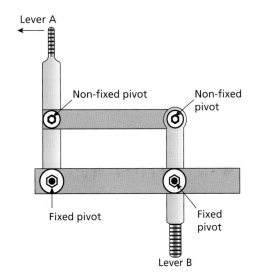

Answer:

_____

QUESTION 5

In the following diagram, pulley 1 turns clockwise. In what direction will pulley 6 turn?

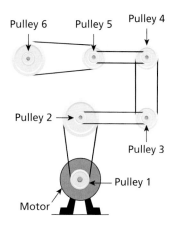

Answer:

_____

QUESTION 6

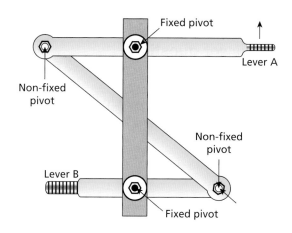

If lever A is pulled up, what will happen to lever B?

Answer:

_____

Unit 20: Reading, Interpreting and Understanding Information in Tables, Charts and Graphs

Section A: Information displayed in tables

Information is often displayed in tables and charts for people to read and retrieve various bits of information. This is common in day-to-day activities such as timetables for buses and trains, viewing sales figures for different products or the results of a questionnaire.

In the electrical industry, tables are often used to record test results or to compare performance data of electrical components, materials and machines. Have a look at the table below, relating to accidents reported to the Health and Safety Executive and answer the questions that follow.

Table ESQCR	Incidents reported under the Electricity Safety, Quality and Continuity Regulations 2002 (as amended) *, 1999/2000 - 2010/2011p.			
Year	**Incident type**			
	Fatal injuries	**Non-fatal injuries**	**Near misses**	**Fires/Explosions**
1999/2000	18	436	2 965	317
2000/2001	13	370	3 346	345
2001/2002	13	383	2 758	218
2002/2003	10	314	2 701	243
2003/2004	9	277	2 738	401
2004/2005	12	280	3 139	350
2005/2006	11	299	4 239	297
2006/2007	20	318	4 045	327
2007/2008	11	300	3 831	211
2008/2009	15	354	4 287	290
2009/2010	12	421	5 120	286
2010/2011	10	570	8 805	404

(Source HSE Aug. 2012). http://www.hse.gov.uk/statistics/tables/index.htm

1. How many categories of reported incidents are shown in the table?

_____

2. In which year were there the most fatal injuries?

_____

3. When were there the most near misses?

_____

4. Has there been an increase or decrease in non-fatal injuries in the last 4 years?

_____

5. Which year had the lowest number of fires/explosions?

_____

Section B: Information displayed in charts

Similar to the use of tables, charts provide a graphical and simplified view of what can sometimes represent a large amount of information, or data.

Column Charts

Using the data from the table in Section A; Incidents reported under the Electricity Safety, Quality and Continuity Regulations, a column chart has been produced to compare the non-fatal accident data from year to year. This gives a pictorial view of how the number of non-fatal accidents changes from year to year.

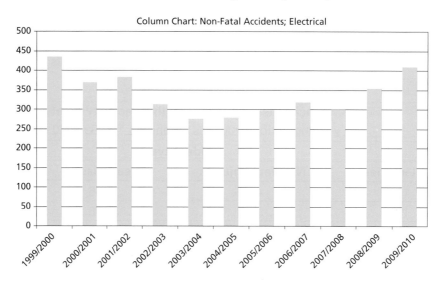

Draw a column chart, using the data from the table in Section A, relating to fatal accidents from 1999 to 2010, using the grid below.

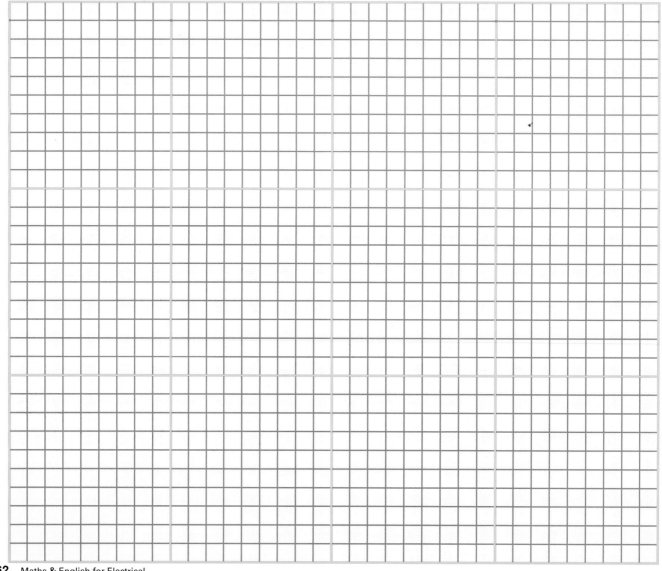

Using the data from the table in Section A; Incidents reported under the Electricity Safety, Quality and Continuity Regulations, a line chart has been produced to compare the non-fatal accident data from year to year. This gives a pictorial view of how the number of non-fatal accidents changes from year to year.

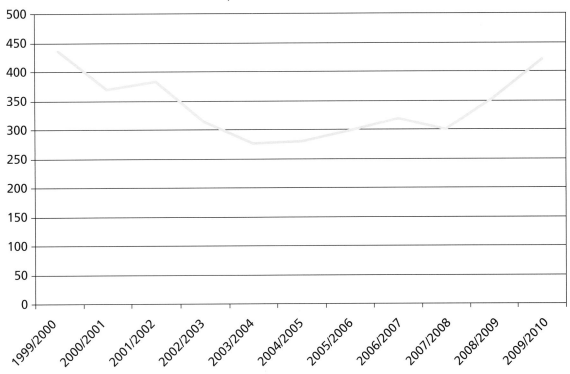

Line Graph: Non-Fatal Accidents; Electrical

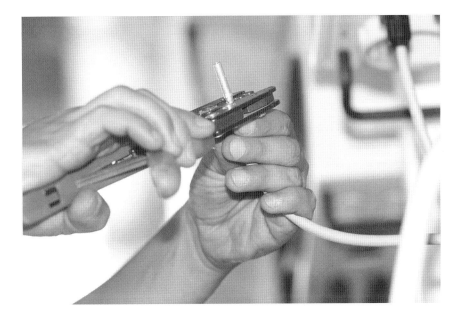

Using the data from the table in Section A, draw a line graph relating to fatal accidents from 1999 to 2010. Use the grid below.

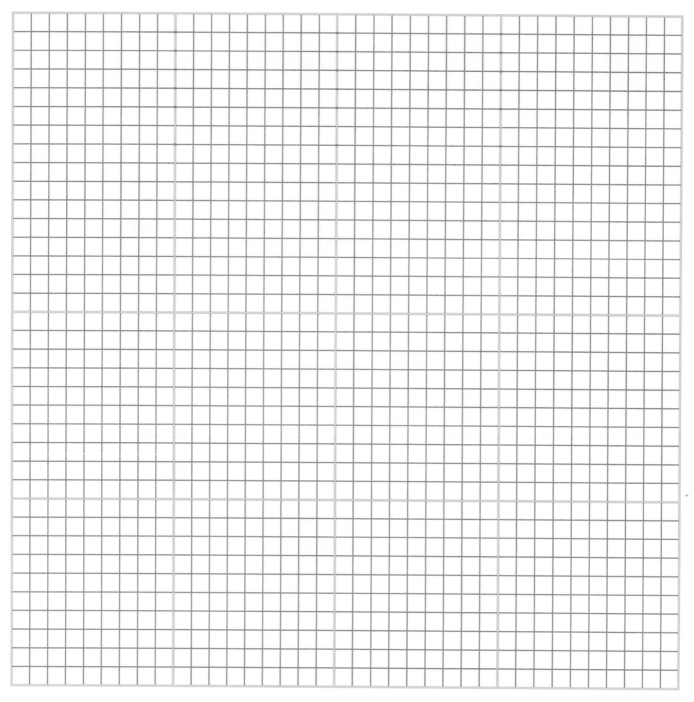

Section C: Gathering and presenting data

Short-answer questions

Specific instructions to students

- This section is designed to help you to both improve your ability to gather, present and interpret data appropriately.
- Read the following questions and answer all of them in the spaces provided.
- You will also need to use extra paper or a computer programme to create a spreadsheet.

Using the internet or catalogues select six different types of (power and hand) tools, suitable for an electrician to use for fixings.

QUESTION 1

Create a spreadsheet to display your list of tools.

(a) Add extra columns and find each of the following:

  a. Approximate price of the tools.

  b. What appropriate fixings the tools can be used for.

  c. What material types and properties (e.g. brickwork) the tools can be used on.

(b) In the space below draw a pie chart using the information in your spreadsheet regarding price of tools:

  a. Tools up to £30.00.

  b. Tools between £31.00 and £75.00.

  c. Tools over £75.00.

QUESTION 2

What conclusions can you draw from your pie chart?

Answer:

_____

_____

_____

QUESTION 3

Describe the variation in the tools that can be used for different fixings. Include the best and worst tools with regards to the number of fixings it can be used on.

Answer:

_____

_____

_____

QUESTION 4

Describe the variation in the tools that can be used for different material types and properties. Include the tool that can be used on the largest number of different materials – which is the most versatile tool?

Answer:

_____

_____

_____

Unit 21:
Practice Written Exam for the Electrical Trade

Reading time: 10 minutes

Writing time: 1 hour 30 minutes

Section A: Literacy

Section B: General Mathematics

Section C: Trade Mathematics

QUESTION and ANSWER BOOK

Section	Topic	Number of questions	Marks
A	Literacy	7	22
B	General Mathematics	11	24
C	Trade Mathematics	40	54
		Total 58	Total 100

The sections may be completed in the order of your choice.

NO CALCULATORS are to be used during the exam.

Spelling

10 marks

Read the passage below and then underline the 20 spelling errors.

A workshop has twulve benches for aprentices to work on. Each apprentise has his or her own toolkit that they had to perchase when they started at the company. Six of the apprentices turned up late on the first day as they could not find the workshop with the addres they were given. The company had moved premices and did not let the new workers know where their new adress was. This caused some confussion initialy, but the apprentices were smart enoufh to ring their supervisor to chek.

Once all of the apprentices had arived, got changed and were at their benches, they were all given diferent tasks to complete. One apprentice had to change five ressistors that were in serieas. Another had to replase three capasitors and several of the others needed to re-soldur circuit boards. One apprentice had to go to the storerom to retreve the soldering irons before they could begin.

Correct the spelling errors by writing them out with the correct spelling below.

_____

_____

Indexing

Put the following words into alphabetical order.

7 marks

Current	Circuit-breaker
Voltage	Capacitor
Power	Resistor
Resistance	Fuse wire
Ohms	Multimeter
Amperes	Cells
Wiring	Volts

_____

_____

_____

_____

_____

_____

Comprehension

Short-answer questions

Instructions to students

- Read the following passage and then answer the questions that follow.

The electrical science teacher welcomed his class, 'good morning everyone, please find a seat and get settled so that we can get the register sorted'.... Register completed the teacher introduced the lesson's objectives.

It was the first lesson with this class and he wanted to introduce the students to the topics of current, voltage, power and resistance.

He began with current, explaining that an electric current is the flow of electric charge. He said that the unit for measuring the sizes of electrical currents is the Ampere, but everyone in the trade referred to them as `Amps'. Most of the students found this interesting, especially because they were hoping to become electricians. The lecturer also pointed out that the common 5 amp or 13 amp fuses will burn out or melt if they are exposed to currents greater than these amounts.

The teacher then moved on to the topic of voltage. He explained that electrical pressure supplied by a cell or generator moves an electric charge through a circuit. This is known as voltage. The unit for measuring voltage is the volt. Normally, a car would use a 12 V battery and a house would use around 230 V.

When the concept of power was introduced, he explained that this was the rate at which electrical appliances used electrical energy. Power is always measured in Watts. He gave an example of a 60 W light bulb being brighter than a 25 W bulb. The reason for this is that a 60 W bulb uses more energy than a 25 W bulb. Some of the energy-efficient bulbs are only 8 W, and despite only using one-sixth of the power, they nevertheless give off similar brightness to a 50 W bulb.

The last topic that the teacher introduced was resistance. He explained that the rate that current flows through a circuit is determined by how much resistance there is. The more resistance there is, the lower the current becomes if the voltage stays the same. He gave a good example of how a 60 W light bulb filament has a lower resistance than a 25 W light bulb filament.

More current can then flow through the 60 W lamp, thus making the light brighter. The unit that resistance is measured in is the Ohm. The lecturer finished off by drawing some symbols and circuits that included these concepts.

QUESTION 1   1 mark

What were the four topics that the lecturer introduced?

Answer:

_____

_____

QUESTION 2   1 mark

In what order did he introduce the topics?

Answer:

_____

_____

QUESTION 3   1 mark

What is the unit of measurement for each topic?

Answer:

_____

_____

QUESTION 4   1 mark

What happens to fuses if they have too much current flowing through them?

Answer:

_____

_____

QUESTION 5   1 mark

Why is a 60 W bulb brighter than a 25 W lamp? How much less power does an energy-efficient globe use?

Answer:

_____

_____

_____

Section B: General Mathematics

QUESTION 1   3 marks

What unit of measurement would you use to measure:

a   Current?

Answer:

_____

b   Resistance?

Answer:

_____

c   Voltage?

Answer:

_____

QUESTION 2   3 marks

Write an example of the following and give an instance of where it may be found in the electrical industry.

a   Percentages

Answer:

_____

b   Decimals

Answer:

_____

c   Fractions

Answer:

_____

QUESTION 3   2 marks

Convert the following units.

a   3 W to kilowatts

Answer:

_____

b   5 kW to watts

Answer:

_____

QUESTION 4   2 marks

Write the following in descending order:

0.7   0.71   7.1   70.1   701.00   7.0

Answer:

_____

QUESTION 5   2 marks

Write the decimal number that is between:

a   0.1 and 0.2

Answer:

_____

b   1.3 and 1.4

Answer:

_____

QUESTION 6   2 marks

Round off the following numbers to two (2) decimal places.

a   5.177

Answer:

_____

b   12.655

Answer:

_____

QUESTION 7   2 marks

Estimate the following by approximation.

a 101×81

Answer:

_____

b 399×21

Answer:

_____

QUESTION 8   2 marks

What do the following add up to?

a   £7, £13.57 and £163.99

Answer:

_____

b   4, 5.73 and 229.57

Answer:

_____

QUESTION 9   2 marks

Subtract the following.

a   196 from 813

Answer:

_____

b   5556 from 9223

Answer:

_____

QUESTION 10   2 marks

Use division to solve:

a $4824 \div 3$

Answer:

_____

b $84.2 \div 0.4$

Answer:

_____

QUESTION 11   4 marks

Using BODMAS, solve:

a $(3 \times 7) \times 4 + 9 - 5$

Answer:

_____

b $(8 \times 12) \times 2 + 8 - 4$

Answer:

_____

Section C: Trade Mathematics

Basic operations

Addition

QUESTION 1   1 mark

An electrician uses 8 m, 22 m, 17 m and 53 m of different types of conduit over 2 months. How much conduit has been used in total?

Answer:

_____

QUESTION 2   1 mark

An electrician charges £227 for labour and £498 for parts. How much is the total bill?

Answer:

_____

Subtraction

QUESTION 1   1 mark

A work van is filled up with 36 L of diesel. The tank is now at its maximum of 52 L. A driver uses the following amounts of diesel on each day:

Monday: 5 L

Tuesday: 11 L

Wednesday: 10 L

Thursday: 8 L

Friday: 7 L

How many litres of diesel are left in the tank?

Answer:

_____

QUESTION 2   1 mark

If an electrician has 224 capacitors in stock and 179 are used over 2 months, how many are left?

Answer:

_____

Multiplication

QUESTION 1   1 mark

An electrician uses two multifunction time relays, four resistors and seven capacitors on one job. How many multifunction time relays, resistors and capacitors would be used on 12 similar jobs?

Answer:

_____

QUESTION 2   1 mark

To connect different electrical parts on the one job, the following wires are used: 2 m of green/yellow wire, 3 m of brown wire and 1 m of blue wire. How much of each wire would be used doing the wiring on 15 similar jobs?

Answer:

_____

Division

QUESTION 1   2 marks

An electrician has a box of 250 resistors.

a How many jobs can be completed if each standard job requires 6 resistors?

Answer:

_____

b   Will any be left over?

Answer:
_____

QUESTION 2   1 mark

If an apprentice electrician earns £288.80 for working a 5-day week, how much is earned per day?

Answer:
_____

Decimals

Addition

QUESTION 1   1 mark

An electrician's tool belt and a hole-saw set are purchased for £17.99 and £56.50 respectively. How much will be paid in total?

Answer:
_____

QUESTION 2   1 mark

An electrician buys a multimeter for £18.75, electrical tape for £6.95, a headlight bulb for £4.95 and a crimping set for £17.50. How much has been spent in total?

Answer:
_____

Subtraction

QUESTION 1   1 mark

An electrician has a 4 L can of de-greaser. It is used on three different electrical cleaning jobs: 1185 mL for job 1, 1560 mL for job 2 and 1135 mL for job 3. How much is left?

Answer:
_____

QUESTION 2   1 mark

An electrician has a 35 m reel of CAT 5 data cable. If 15.48 m is used on one job, 12.76 m is used on another and 3.44 m is used on the last job, how much is left on the reel?

Answer:
_____

Multiplication

QUESTION 1   1 mark

An apprentice replaces six fluorescent light tubes at a cost of £6.99 each and four light switches at a cost of £11.99 each. What is the total value of the goods?

Answer:
_____

QUESTION 2   1 mark

A packet of 100 single screw electrical connectors costs £38.50. If an electrician uses six packets, how much is the total cost?

Answer:
_____

Division

QUESTION 1   1 mark

An electrician takes 12 hours to complete three jobs around a house. His total labour bill amounts to £582.48. How much does he earn per hour?

Answer:
_____

QUESTION 2   1 mark

A wholesaler buys 240 electrical connectors in bulk at a total cost of £3600. How much is the cost of one electrical connector?

Answer:
_____

Fractions

QUESTION 1   1 mark

$\frac{2}{3} + \frac{3}{4} =$

Answer:

QUESTION 2 1 mark

$$\frac{4}{5} - \frac{1}{3} =$$

Answer:

_____

QUESTION 3 1 mark

$$\frac{2}{3} \times \frac{1}{4} =$$

Answer:

_____

QUESTION 4 1 mark

$$\frac{3}{4} \div \frac{1}{2} =$$

Answer:

_____

Percentages

QUESTION 1 2 marks

An electrical repair bill comes to £2546.00.

a   How much is 10% of the bill?

Answer:

_____

b   What is the final bill once the 10% is taken off?

Answer:

_____

QUESTION 2 2 marks

An apprentice buys an electrician's manual on DVD and a set of screwdrivers. The total comes to £37.80.

a   How much is 10% of the bill?

Answer:

_____

b   What is the final bill once the 10% is taken off?

Answer:

_____

Voltage, current and resistance

Voltage

QUESTION 1 1 mark

What is the voltage in a circuit that has a current of 5 A and a resistance of 10 Ω?

Answer:

_____

QUESTION 2 1 mark

What is the voltage in a circuit that has a current of 15 A and a resistance of 16 Ω?

Answer:

_____

Current

QUESTION 1 1 mark

What is the current in a circuit with a voltage of 6 V and a resistance of 3 Ω?

Answer:

_____

QUESTION 2 1 mark

What is the current in a circuit with a voltage of 240 V and a resistance of 6 Ω?

Answer:

_____

Resistance

QUESTION 1 1 mark

What is the resistance of a circuit with a voltage of 12 V and a current of 2 A?

Answer:

_____

QUESTION 2   1 mark

What is the resistance of a circuit with a voltage of 240 V and a current of 12 A?

Answer:

_____

Measurement conversions

QUESTION 1   1 mark

How many millimetres are there in 3.85 m?

Answer:

_____

QUESTION 2   1 mark

How many metres does 2285 mm convert into?

Answer:

_____

Measurement – length and area

Circumference

QUESTION 1   2 marks

Find the circumference of a warehouse light fitting with a diameter of 800 mm.

Answer:

_____

QUESTION 2   2 marks

Find the circumference of a pulley with a diameter of 150 mm.

Answer:

_____

Area

QUESTION 1   2 marks

An electrical workshop measures 45 m by 12 m. What is the total area?

Answer:

_____

QUESTION 2   2 marks

A light switch cover is 115 mm by 105 mm. What is the total area?

Answer:

_____

Earning wages

QUESTION 1   2 marks

A first-year electrical apprentice earns £280.60 net (take home) per week. How much does she earn per year? (Note that there are 52 weeks in a year.)

Answer:

_____

QUESTION 2   2 marks

A control circuit on a farm grain dryer is damaged by vermin. The labour bill comes to £2860. If the electrician spends 70 hours working on the repairs, how much per hour does he charge?

Answer:

_____

Squaring numbers

QUESTION 1   2 marks

What is 9^2?

Answer:

_____

QUESTION 2   2 marks

A workshop has an area for a hoist that is 8.2 × 8.2 m.
What is the total area?

Answer:

_____

Ratios

QUESTION 1   2 marks

A driver cog has 20 teeth and the driven cog has 60
teeth. What is the ratio, in the lowest form, of the driver
cog to the driven cog?

Answer:

_____

QUESTION 2   2 marks

The ratio of transformer windings is 8:3. If winding A
has 40 turns, how many turns are on winding B?

Answer:

_____

Mechanical reasoning

QUESTION 1   1 mark

Pully 1 and pully 2 each measure 50 mm across
their diameters. Pully 3 measures 100 mm across the
diameter. How many times will pulleys 1 and 2 turn if
pully 3 turns three times?

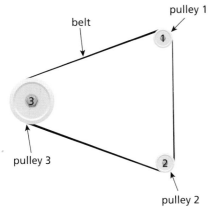

_____

QUESTION 2   1 mark

Each cog in the below diagram has 16 teeth and
they interlock with one another. If cog 5 turns in an
anticlockwise direction, which way will cog 1 turn?

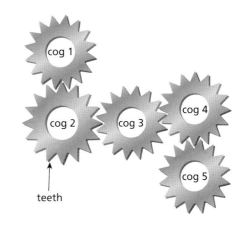

_____

Electrical Glossary

Access equipment   Equipment used to access the work area or provide a platform to work from, usually when working at height, e.g. ladders, stepladders, scaffolding, scaffold towers, scissor lifts, etc.

Alternating current (AC)   An electric current that reverses direction at regular recurring intervals

Ampere (amp)   A unit of electric current

Bus bar   A heavy copper bar used as the primary power source to carry heavy currents. A bus bar can also be used to make a common connection between several circuits

Cable   Generally wire used for conducting electricity that is covered with some type of insulation

Capacitor   An electrical device that consists of two conducting surfaces that are oppositely charged and that are separated by a thin layer of insulation material

Circuit   A system of conducting parts through which an electrical current flows

Circuit-breaker   A device that automatically opens a circuit should the current exceed the rated amount. The circuit-breaker can also be re-set

Conductor   Any material, (usually metal), intended to transmit or carry an electric current

Conduit   Metal or plastic pipe used to provide mechanical protection for single core conductors

Contactor   An electro-magnetic switch

Contractor   A person that works on a contract basis. Many people within the building industry work as contractors

Current (I)   The transfer of electric charge through a conductor. It is measured in amperes

Cycle   A complete positive and a complete negative alternation of voltage or current

Direct current (DC)   An electric current in which there is a non-stop transfer of charge in one direction

Farad   The unit of measurement for capacitance

Flex   Flexible type of cable

Frequency   The number of complete cycles in a unit of time. The unit of measurement for frequency is hertz

Fuse   An electrical device of wire or a strip of fusible metal that can melt and therefore open a circuit if the current exceeds the rated amount. A safety device that destroys itself and is replaced with a new fuse

Horsepower   A unit of power equal to 746 W of electrical power

Installation   A fixed wiring system or the process of fixing (installing), a wiring system

Insulation   The non-conducting covering of a current carrying conductor. A material that acts as an insulator to electric current

Insulator   A material that has high resistance to the flow of an electric current

Joule   A unit of electrical energy

Junction box   A box for inserting and joining cables or wires

Kilowatt (kW)   A unit of electrical power equal to 1000 W

Kirchoff's law   The sum of the currents entering a junction is equal to the sum of the currents leaving that junction

Luminaire   A lamp and lamp fitting that includes the control gear necessary for its operation

Ohm   A unit of measurement for resistance

Ohm's law   Current is directly proportional to voltage and inversely proportional to resistance

Parallel circuit   A method of connecting a circuit so that the current has two or more paths to follow

Power   The rate of doing work. The unit for power is watts

PVC   Polyvinyl chloride, a common insulation material

Resistance (R)   The opposition a material gives to the flow of electrons. It is measured in ohms

Transformer   An electrical machine, without moving parts, used to step voltage and current up or down

Trunking   Metal or plastic rectangular tube used to provide mechanical protection for single core conductors

Volt (V)   A unit of electrical potential or pressure

Voltage   The electromotive force or electrical pressure that is measured in volts

Voltage drop   The potential difference measured across current-limiting elements in a circuit

Watt (W)   A unit of measure of power

Wavelength   The distance travelled by a wave during the time interval covered by a cycle

Wholesaler   A trade supplier or distributor of electrical equipment, materials and tools

Maths and English Glossary

Adjectives   Describes things, people and places, such as 'sharp', 'warm' or 'handsome'

Adverbs   Describes the way something happens, such as 'slowly', 'often' or 'quickly'

Homophones   Words that sound the same, but are spelt differently and have different meanings

Imperial   A system of units for measurements e.g. pounds and inches

Metric   An international system of units for measurement. This is a decimal system of units based on the metre as a unit length and the kilogram as a unit mass

Nouns   Names of things, people and places, such as 'chair', 'George' or 'Sheffield'

Pronouns   Short words like 'it', 'you', 'we' or 'they', etc. used instead of actual names

Ratio   A way to compare the amounts of something

Verbs   Words to describe what you are 'doing', such as 'to mix', 'smile/frown' or 'walking'

Formulae and Data

Circumference of a Circle

$C = \pi \times d$
where: C = circumference, π = 3.14, d = diameter

Diameter of a Circle

Diameter (d) of a circle $= \dfrac{\text{circumference}}{\pi(3.14)}$

Area

Area = length $\times$ breadth and is given in square units
$\quad A = l \times b$

Voltage

Voltage (V) = current (I) $\times$ resistance (R)
$\quad\quad V = I \times R$

Current

Current (I) $= \dfrac{\text{voltage } (V)}{\text{resistance } (R)}$

$\quad\quad I = \dfrac{V}{R}$

Resistance

Resistance (R) $= \dfrac{\text{voltage } (V)}{\text{current } (I)}$

$\quad\quad R = \dfrac{V}{I}$

Times Tables

1
1 × 1 = 1
2 × 1 = 2
3 × 1 = 3
4 × 1 = 4
5 × 1 = 5
6 × 1 = 6
7 × 1 = 7
8 × 1 = 8
9 × 1 = 9
10 × 1 = 10
11 × 1 = 11
12 × 1 = 12

2
1 × 2 = 2
2 × 2 = 4
3 × 2 = 6
4 × 2 = 8
5 × 2 = 10
6 × 2 = 12
7 × 2 = 14
8 × 2 = 16
9 × 2 = 18
10 × 2 = 20
11 × 2 = 22
12 × 2 = 24

3
1 × 3 = 3
2 × 3 = 6
3 × 3 = 9
4 × 3 = 12
5 × 3 = 15
6 × 3 = 18
7 × 3 = 21
8 × 3 = 24
9 × 3 = 27
10 × 3 = 30
11 × 3 = 33
12 × 3 = 36

4
1 × 4 = 4
2 × 4 = 8
3 × 4 = 12
4 × 4 = 16
5 × 4 = 20
6 × 4 = 24
7 × 4 = 28
8 × 4 = 32
9 × 4 = 36
10 × 4 = 40
11 × 4 = 44
12 × 4 = 48

5
1 × 5 = 5
2 × 5 = 10
3 × 5 = 15
4 × 5 = 20
5 × 5 = 25
6 × 5 = 30
7 × 5 = 35
8 × 5 = 40
9 × 5 = 45
10 × 5 = 50
11 × 5 = 55
12 × 5 = 60

6
1 × 6 = 6
2 × 6 = 12
3 × 6 = 18
4 × 6 = 24
5 × 6 = 30
6 × 6 = 36
7 × 6 = 42
8 × 6 = 48
9 × 6 = 54
10 × 6 = 60
11 × 6 = 66
12 × 6 = 72

7
1 × 7 = 7
2 × 7 = 14
3 × 7 = 21
4 × 7 = 28
5 × 7 = 35
6 × 7 = 42
7 × 7 = 49
8 × 7 = 56
9 × 7 = 63
10 × 7 = 70
11 × 7 = 77
12 × 7 = 84

8
1 × 8 = 8
2 × 8 = 16
3 × 8 = 24
4 × 8 = 32
5 × 8 = 40
6 × 8 = 48
7 × 8 = 56
8 × 8 = 64
9 × 8 = 72
10 × 8 = 80
11 × 8 = 88
12 × 8 = 96

9
1 × 9 = 9
2 × 9 = 18
3 × 9 = 27
4 × 9 = 36
5 × 9 = 45
6 × 9 = 54
7 × 9 = 63
8 × 9 = 72
9 × 9 = 81
10 × 9 = 90
11 × 9 = 99
12 × 9 = 108

10
1 × 10 = 10
2 × 10 = 20
3 × 10 = 30
4 × 10 = 40
5 × 10 = 50
6 × 10 = 60
7 × 10 = 70
8 × 10 = 80
9 × 10 = 90
10 × 10 = 100
11 × 10 = 110
12 × 10 = 120

11
1 × 11 = 11
2 × 11 = 22
3 × 11 = 33
4 × 11 = 44
5 × 11 = 55
6 × 11 = 66
7 × 11 = 77
8 × 11 = 88
9 × 11 = 99
10 × 11 = 110
11 × 11 = 121
12 × 11 = 132

12
1 × 12 = 12
2 × 12 = 24
3 × 12 = 36
4 × 12 = 48
5 × 12 = 60
6 × 12 = 72
7 × 12 = 84
8 × 12 = 96
9 × 12 = 108
10 × 12 = 120
11 × 12 = 132
12 × 12 = 144

Multiplication Grid

×	1	2	3	4	5	6	7	8	9	10	11	12
1	1	2	3	4	5	6	7	8	9	10	11	12
2	2	4	6	8	10	12	14	16	18	20	22	24
3	3	6	9	12	15	18	21	24	27	30	33	36
4	4	8	12	16	20	24	28	32	36	40	44	48
5	5	10	15	20	25	30	35	40	45	50	55	60
6	6	12	18	24	30	36	42	48	54	60	66	72
7	7	14	21	28	35	42	49	56	63	70	77	84
8	8	16	24	32	40	48	56	64	72	80	88	96
9	9	18	27	36	45	54	63	72	81	90	99	108
10	10	20	30	40	50	60	70	80	90	100	110	120
11	11	22	33	44	55	66	77	88	99	110	121	132
12	12	24	36	48	60	72	84	96	108	120	132	144

Maths and English for Electrical
Online Answer Guide

To access the Answer Guide for Maths and English for Electrical follow these simple steps:

1) Copy the following link into your web browser:

http://www.cengagebrain.co.uk/shop/isbn/9781408077535

2) Click on the Free Study Tools Link.

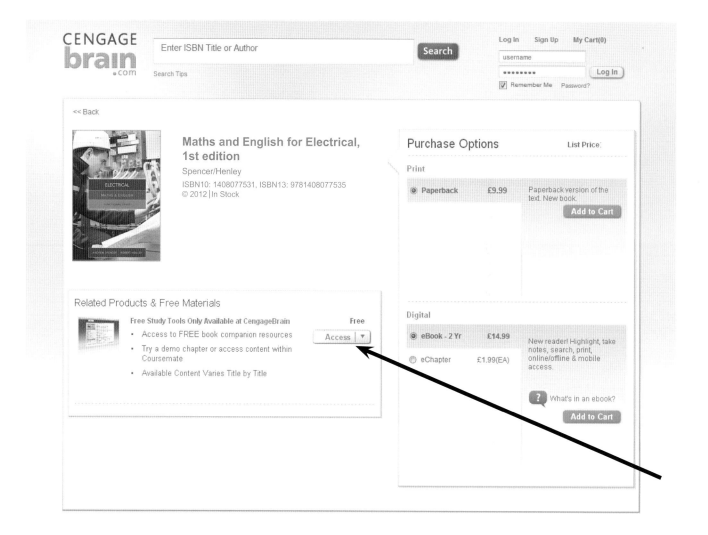

Download each answer section by selecting the chapter tab and clicking on the Answer Guide **under 'Section'**

Download the whole workbook by selecting the chapter tab and clicking on the Answer Guide **under 'Book Resources'**

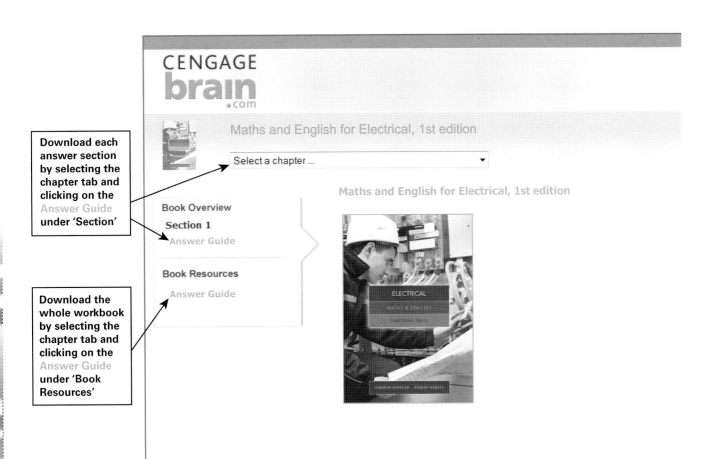

CENGAGE
brain.com

Maths and English for Electrical, 1st edition

Select a chapter ...

Maths and English for Electrical, 1st edition

Book Overview

Section 1
Answer Guide

Book Resources
Answer Guide

Notes

Notes

Notes

Notes

Notes

Notes

Notes